ADVANCED PHYSICS - II
Enhanced with Bijective Analysis

Amrit Srecko Sorli

The foreword by the author

20th-century physics is based on the belief that fundamental arena of the universe is space-time where time is 4th dimension of space. Particles and massive bodies are moving in an empty space-time deprived of physical properties. This false picture is the main obstacle for the advance of physics.

In this book, a new vision based on bijective research methodology is proposed where each physical equation satisfies the bijective function i.e. each element of an equation corresponds to a particular element in the real world. Likewise, every equation and physical model has bijective correspondence with the real world. Bijective analysis realizes Einstein's vision of completeness of a physical theory. According to Einstein a theory can be considered complete if every element of physical theory has a counterpart in the physical reality.

One will be surprised by seeing that some of today's leading physical models have no bijective correspondence with real world. Modern physics is both predicting and discovering physical phenomena theoretically, which sometimes has no correspondence with the physical world. Nobel Prize was delivered for the discovery of Higgs field which is pure theoretical failure. A bunch of articles were published in which equations are in mixed condition, having no understanding of the real meaning of it. Nicola Tesla was right by saying, '*Today's scientists have substituted mathematics for experiments, and they wander off through equation after equation, and eventually build a structure which has no relation to reality*'.

Physics necessarily requires bijective analysis in order to keep the actual bonding with the real world. It is time to build the solid foundations of physics, where the theoretical model and the real world are associated with the bijective function. This book is the first step in this important task.

Table of content:

Bijective Analysis of Physical Equations and Physical Models

Euclidean-Planck metrics of space, particle physics and cosmology

Minkowski space-time and Einstein's Now conundrum

Advanced Relativity: Reintroduction of absolute Frame of Reference

Epistemology Crisis of Today's Physics

Advanced Relativity for the Renaissance of Cosmology and Evolution of Life

Advanced Relativity: Unification of Space, Matter and Consciousness

Advanced Relativity: Multidimensionality of Consciousness and Mind, Origin of Life, PSI Phenomena

Experiential Methodology in Consciousness Research

Unified Field Theory Based on Bijective Methodology

Bijective Analysis of Physical Equations and Physical Models

Abstract

A bijective function $f: X \to Y$ is a function between the elements of two sets, where each element of one set is paired with exactly one element of the other set, and where each element of the other set is paired with exactly one element of the first set. We can define every physical equation as the set X and the corresponding physical reality that the equation describes as the set Y. Every element in a given equation is an element in the set X, and each element in set X should have exactly one paired element in the set Y. The bijective analysis confirms that Newtonian physics satisfies the bijective function. On the other hand, not all of the equations in the Theory of Relativity satisfy the bijective function. Additionally, the Higgs mechanism does not satisfy the bijective function.

Key words: bijective function, Newton physics, Theory of Relativity, gravitational waves, Higgs mechanism, holographic mass, proton's Schwarzschild radius

1. Introduction

In mathematics, bijective function, is a function between the elements of two sets, where each element of one set is paired with exactly one element of the other set, and each element of the other set is paired with exactly one element of the first set. There are no unpaired elements:

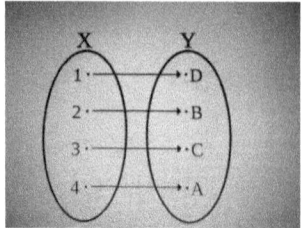

Figure 1: A bijective function, $f: X \rightarrow Y$, where set X is {1, 2, 3, 4} and set Y is {A, B, C, D}. For example, $f(1) = D$

Let's analyse one of the most fundamental equations with the help of bijective function:

$$d = v \cdot t \quad (1),$$

where d is distance, v is velocity, and t is time as the duration of motion on distance d.

In equation (1), every element of set X corresponds to exactly one element in the set Y. Between equation (1) and physical reality there is a one-to-one bijective correspondence with physical reality. Equation (1) is a 100% exact picture of physical reality. The same is valid for all equations of Newtonian physics.

In order to rescue the ether model, Lorentz developed a formula for length contraction:

$$L = \frac{L_0}{\gamma} \quad (2),$$

where L_0 is proper length, L is the length of a moving object and γ is the Lorentz factor.

Length contraction has never been observed nor measured in physical reality. We do not have any experimental data that length contraction actually exists in physical reality. Between formula (2) and physical reality there is no bijective correspondence. We cannot assume that formula (2) is a 100% exact picture of physical reality. Lorentz did not explain, (nor has anyone ever explained) the physical circumstances that would cause length contraction. For more than 100 years formula (2) has been accepted in physics as if it had bijective correspondence with physical reality, even though it does not.

2. Space-time model has no bijective correspondence with physical reality

For 100 years in physics it has been fully accepted that time is the 4th dimension of space. However, bijective analysis confirms that time is not the 4th dimension of space; rather, time is the duration of motion in space (Fiscaletti, Sorli, 2015). Thus, the element of »coordinate time« in Special Relativity has no bijective correspondence with physical reality. Likewise, the formula (3) below has no bijective correspondence with physical reality:

$$X_4 = ict \quad (3),$$

where X_4 is meant to be the 4th dimension of space, i is an imaginary number, c is light speed, and t is time as duration. In order to make the formula more "realistic" physicists removed the i and the formula became:

$$X_4 = ct \quad (4).$$

Formula (4) is equal to formula (1) and has bijective correspondence with physical reality if X_4 represents distance in space. But physicists were convinced that X_4 is the time coordinate, and so they replaced light speed c (which is constant) with the number 1 and they got:

$$X_4 = t \quad (5).$$

In formula (5) time has become distance, which has no bijective correspondence with physical reality.

Bijective analysis confirms that the Minkowski manifold has no bijective correspondence with physical reality. We are not living in space and time, we are living only in space. The universe exists only in space, and time, when measured, is the duration of motion in space. In space is always NOW (Sorli et al., 2017b).

3. Time dilation has bijective correspondence with physical reality

GPS systems confirm that clocks on satellites have a slower rate than the clocks on the Earth's surface by 7 microseconds per day (Real-World Relativity, 2017). The formula that describes time dilation has bijective correspondence with physical reality:

$$t = t_0 \cdot \gamma \quad (6),$$

where t is the time on the satellite, t_0 is the time on the Earth's surface and γ is the Lorentz factor. Time dilates in the sense that the rate of the clocks and the velocity of all physical changes becomes slower. Formula (6) does not mean that time, as the fourth coordinate of space, could dilate so that light would then need more time to pass the distance from point A to point B in space. The Shapiro experiment, which confirms that in a strong gravitational field light speed is slightly diminished, is explained as "gravitational time dilation," in the sense that time, as the 4th coordinate of space, is conceived to be dilating in the relatively stronger gravitational field, which then causes light to need more time to move from A to B. However, this imagined "gravitational time dilation" has no bijective correspondence with physical reality. What really happens is that in a relatively stronger gravitational field the permittivity and permeability of space changes, which changes then minimally diminish the velocity of light:

$$c = \frac{1}{\varepsilon_0 \mu_0} \quad (7),$$

where ε_0 is electric permittivity and μ_0 is magnetic permeability. The minimal variability of light speed does not contradict Special Relativity; it only confirms that gravity has an influence on light speed.

4. **Dilation and contraction of space have no bijective correspondence with physical reality**

General Relativity predicts that space can be dilated and contracted by gravitational waves (Tiec, Novak, 2017), which were discovered in (Abott et al., 2016). However, the idea that a gravity wave could shrink and dilate space and therefore affect the beams of the LIGO interferometer, has no bijective correspondence with the physical world. Dilation and contraction of space have never been directly observed. What was observed in LIGO is that laser light has minimal time dilation and minimal time contraction. It is not that the beams of the interferometer are changing length; rather, what actually happens in LIGO is that the gravity wave changes the permittivity and permeability of space which in turn then causes the minimal variations of laser light speed as we have seem in chapter 3.

5. Higgs mechanism has no bijective correspondence with physical reality

The Higgs mechanism has its origin in supersymmetry (SUSY), which is an extension of the Standard Model: "Supersymmetry predicts a partner particle for each particle in the standard model, to help explain why particles have mass. At first sight, the Standard Model seems to predict that all particles should be massless, an idea at odds with what we observe around us. Theorists have come up with a mechanism to give particles masses that requires the existence of a new particle, the Higgs boson" (CERN, 2018). There is no experimental data that proves that SUSY has bijective correspondence with physical reality. Additionally, the model of the Higgs field has no bijective correspondence with physical reality. At CERN they discovered that the Higgs boson is a momentary characteristic flux of energy with life time $1{,}56 \cdot 10^{-22}\,s$ released by the collision of two protons. Frankly, bijective analysis confirms that the existence of the Higgs boson does not prove the existence of the Higgs field. Between the discovery of the Higgs boson and the existence of the Higgs field there is a huge "bijective gap."

The assumption of Higgs mechanism: "Must be a field which gives mass to the particles", seems not right. The right assumption is: "Must be a field which gives inertial mass to the particles". In this article it is shown that the field that gives inertial mass to the particles is space itself, namely, its variable energy density. Inertia of elementary particles which fully respects mass-energy equivalence is solved with the introduction of Euclidean-Planck metrics of space (EPM) (Sorli et al., 2018). In EPM the inertia of a given massive particle, massive body, or stellar object has its origin in the diminished energy density of space ρ_{SE} in the centre of a given physical object, which causes outer higher pressure of space with Planck energy density ρ_{PE} towards the centre of given physical object:

$$\rho_{SE} = \rho_{PE} - \frac{mc^2}{V} \quad (8)\,[7],$$

where m is the mass of the object, c is the light speed and V is its volume.

We can rearrange formula (8) and we get:

$$m = \frac{(\rho_{PE} - \rho_{SE}) \cdot V}{c^2} \quad (9).$$

The inertial mass m_i of a given massive particle with mass m has its origin in the diminished energy density of space in its centre. The value of inertial mass m_i is equal to the value of its mass m:

$$m_i = m = \frac{(\rho_{PE} - \rho_{SE}) \cdot V}{c^2} \quad (10).$$

Actually we define the amount of mass m of a given physical object by measuring its inertial mass m_i.

Formula (9) shows a given physical object with mass m diminishes the energy density of space exactly according to the amount of its energy E. We can express this by multiplying the formula by c^2 and we get:

$$E = mc^2 = (\rho_{PE} - \rho_{SE}) \cdot V \quad (11).$$

The right part of equation (11) has bijective correspondence with physical reality, because the idea of 20th century physics that space is empty and deprived of physical properties is false. Bijective analysis confirms that matter is energy and that matter exists in space, which means that space also is energy (Sorli et al., 2018). NASA research confirms that universal space has a Euclidean shape (NASA 2014), which means that the curvature of space in General Relativity is a mathematical description of variable energy density of space (Fiscaletti, Sorli, 2014), which is its real physical property. Equation (11) is in accord with Einstein view on mass and energy: "Mass of a body is a measure of its energy-content" (Einstein 1916).

6. **Quarks, Higgs boson, W and Z bosons are not "particles" as are protons, electrons, and photons**

Particle physics research methodology for the last fifty years has primarily involved the scattering and collision of known particles, for example protons, as a way of search for previously unknown particles. Some of these previously unknown particles have extremely short lifetimes, which calls into question whether or not these particles have any real physical existence outside the context of the experimental environment of colliders. In this chapter we will challenge the objective existence of quarks, W and Z bosons.

The research methodology of scattering and collision generally involves shooting particles (for example protons) at targets (or between them in LHC), thereby smashing them and then observing the new particles that are created. These newly created particles have extremely short lifetimes. For example, the top quark lifetime is $5 \cdot 10^{-25} s$, the W and Z boson lifetimes are about $3 \cdot 10^{-25} s$, and the Higgs boson lifetime is $1,56 \cdot 10^{-22} s$. In our scientific imagination or conception, the term "subatomic particle" is usually associated with some element (a quark for example) that is a consistent part of a particle, such as a proton, that is itself extremely stable. The question arises: do quarks have an existence on their own? Put another way, are quarks real elements of the proton, or are they only characteristic fluxes of energy, or a particular pattern of energy, that is consistently released by scattering and collision? In this chapter this question will be examined in some detail.

A proton is a stable particle the lifetime of which is $2,1 \cdot 10^{29} s$. The existent description of the proton in the Standard Model is known: a proton is composed of two up quarks and one down quark which are bound together with gluons. When we "smash" the proton we observe quarks as its composite parts. We will present an alternative model of the proton as a vortex of space energy in the form of a torus where space energy is circulating in a closed torus loop. Proton energy circulation creates its electric positive charge and magnetic momentum. The idea that the torus is the fundamental form in the universe is becoming actual in scientific thought (Meijer, Gesing 2016; Meijer, Gesing 2017). The idea that the proton could be in the form of torus is also entering physics (Piskounova 2017). A proton as the vortex of space energy in torus form has no composite parts; a photon is one unit, which when smashed falls apart in characteristic fluxes of energy which we observe as quarks. In the proton itself, there are no quarks and there are no gluons which would bind together quarks. Gluons are characteristic "sparkles" of energy released when a proton is smashed; they are not its composite parts.

A neutron, when not helping to compose the nucleus of an atom, is unstable, with a lifetime of $881,5 \pm 1,5 s$. A neutron decays into a proton and electron. The conception that the neutron is a consistent part of the nucleus has no bijective correspondence with physical reality, because a neutron alone is unstable and is composed of a proton and electron. A neutron is not particle in the sense of having a "stable life time," as for example a proton, or a photon which life time is about 10^{18} $years$, or an electron which life time is about $6,6^{28}$ $years$. Proton, electron and photon are particles in the sense of being composite parts of atoms. On the other hand, gluons and quarks seem not to be "particles" in the sense of being composite parts of atoms.

From this perspective W bosons in Z bosons which life time is about $3 \cdot 10^{-25} s$ are also not "particles," as are the proton, electron and photon. Rather, W bosons in Z bosons are momentary fluxes of energy characteristic of the weak interaction.

In general bijective analysis of Higgs boson, W and Z bosons and gluons, all of which have extremely short lifetimes, confirms that these physical entities are momentary energy fluxes and cannot be considered as "particles," which are constitutive elements of atoms.

Recent research speculates that the Higgs boson, which constitutes the Higgs field, appears before inflation (according to the Big Bang theory) and that the coupling of the Higgs field with gravitation causes accelerated expansion of the universe (Brezukov 2014). Our research confirms that the radius of the universe in the Big Bang model is too small for the rate 10^4 to be placed in existent mapped universe and that the inflation signal would still be 32,9 billion light years distant from the planet Earth (Sorli et al., 2018). The Big Bang model has no scientific validity and also weakens the validity of the existence of the Higgs field, which is also extremely weak from the point of epistemology (Sorli, Kaufman, 2018).

7. Variable energy density of space and vortex model of elementary particles

In this article a model of the proton is presented in which the proton, as a vortex of space, is the only particle which has inertial mass. In the centre of the proton vortex the energy density of space ρ_{SE} is lower, and that causes the outside pressure of space where energy density has the Planck value ρ_{PE}. Inertial mass m_i of the proton is equal to its mass in the Einsteinian sense: "Mass of a body is a measure of its energy-content" (Einstein 1916). Einstein has proved in his General Theory of Relativity that inertial mass m_i and gravitational mass m_g of a given physical object are equal. The formula below which is derived from (9), clearly express the relation between proton mass, proton inertial mass, and proton gravitational mass:

$$m_i = m_g = m = \frac{(\rho_{PE} - \rho_{SE}) \cdot V}{c^2} \quad (12).$$

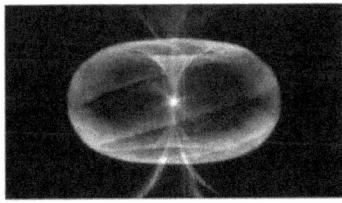

Figure 2: Proton is a vortex of space energy in the form of torus

When the proton is accelerated in a cyclotron, proton gain its relativistic energy E_R which we can calculate according to the following formula derived from the formula (11):

$$E_R = \gamma \cdot mc^2 = \gamma \cdot (\rho_{PE} - \rho_{SE}) \cdot V \quad (13),$$

where γ is the Lorentz factor. With the increase of speed, the proton vortex interacts with the energy of space in the way that it absorbs the additional energy of space; we can call this interaction "dragging effect" between relativistic proton and space. The fact that relativistic proton interacts with space opens the possibility for the reinterpretation of the "absolute frame of reference" which will be carefully examined.

Calculation of diminished energy density of space ρ_{SE} in the centre of the proton vortex in empty space far away from stellar objects is following:

Planck energy density ρ_{PE}: $4.633 \times 10^{113} \, J/m^3$

Mass of the proton m_p: $1.6727 \times 10^{-27} \, kg$

Volume of the proton: $V_p = 2.5 \times 10^{-45} \, m^3$

Formula for energy density of space ρ_{SE} in the centre of proton vortex we derive from formula (12):

$$\rho_{SE} = \rho_{PE} - \frac{m_p \cdot c^2}{V} \quad (14).$$

cf

$$\rho_{SE} = 4.633 \times 10^{113} \, J/m^3 - \frac{(1.6727 \times 10^{-27} \, kg)(8.99 \times 10^{16} \, m^2 s^{-2})}{2.5 \times 10^{-45} \, m^3}$$

$$\rho_{SE} = 4.633 \times 10^{113} \, J/m^3 - \frac{15.04 \times 10^{-11} \, J}{2.5 \times 10^{-45} \, m^3}$$

$$\rho_{SE} = 4.633 \times 10^{113} J/m^3 - 6.015 \times 10^{34} J/m^3$$

$$\Delta_{space.energy.density} = \rho_{PE} - \rho_{SE} = 6.015 \times 10^{34} J/m^3$$

In the centre of a proton the vortex energy density of interstellar space is smaller than the outer Planck energy density of space of $6,015 \cdot 10^{34} J/m^3$. Protons are continuously created throughout universal interstellar space out of space energy. In black holes the energy density of space is at the minimum, which causes proton to become unstable and disintegrate back into the energy of space. The value of the energy density of space in the centre of a black hole with the mass of the Sun is smaller than in outer intergalactic space by $1,582 \cdot 10^{36} J/m^3$ (Sorli et al., 2018). The idea that a proton could be a kind of mini black hole [16] is not in accord with the calculations above and disagree with regard the energy density of space in the centre a of proton vortex and in the centre of a black hole for the value:

$$\Delta_{BLACKHOLE.CENTRE-PROTON.CENTRE} = 1.582 \times 10^{36} - 6.015 \times 10^{34} J/m^3$$

$$\Delta_{BLACKHOLE.CENTRE-PROTON.CENTRE} = 1.522 \times 10^{36} J/m^3,$$

which is an error of magnitude 10^{36}. Building a model of mass of the proton on the proposition that proton is a mini black hole, leads to conclusions that do not seem to have correspondence with the physical reality of the proton, namely: "The mass of the proton can be described as the exchange of information across the boundary of its Event Horizon, and that its gravitational mass is equivalent with the strong force when special relativity and mass dilation is considered" (Haramein, 2013). Schwarzschild radius of the proton is mathematical construct with no correspondence to any physical reality. We can calculate Schwarzschild radius for the Sun which is 3000 meters, but we have to understand that this calculation has no physical meaning, because Sun will never develop into a black hole. Recent research confirms Sun will develop into red giant and further on in white dwarf (Gesicki at al., 2018). The calculation of Schwarzschild radius of Sun just helps us to understand theoretically how a star as our Sun has to be compressed in order to reach the diminished energy density of the space in its centre which is characteristic for a black hole (see figure 4 in this chapter). Recent research confirms that for a star to become a black hole 2,16 solar masses is needed. If the star has less mass than 2,16 solar masses it will develop into a neutron star (Rezzolla et al., 2108). The famous formula for Schwarzschild radius was developed by German astronomer Karl Schwarzschild back in 1916:

$$r_s = \frac{2GM}{c^2} \qquad (15),$$

where r_s is Schwarzschild radius, G is gravitational constant, and M is the mass of a star. Discovery of Rezzolla confirms that formula (15) has physical meaning when condition (16) is fulfilled:

$$M \succ 2,16 \cdot M_{Sun} \qquad (16).$$

Bijective analysis confirms that the model of proton as a black hole has no bijective correspondence with the physical world. Regarding the idea of proton having Schwarzschild radius as something that has physical existence, the famous quote of Nicola Tesla is fully justified: "Today's scientists have substituted mathematics for experiments, and they wander off through equation after equation, and eventually« build a structure which has no relation to reality".

We can develop formula (12) in the following formula:

$$\rho_{SE} = \rho_{PE} - \frac{3m \cdot c^2}{4\pi \cdot (r+d)^3} \quad (16),$$

where m is the mass of the material object, r is radius of the material object and d is the distance from the centre of the material object to a given point T (see figure 3 below). When $d=0$, one gets energy density of space in the centre of the stellar object. When $d \succ 0$, one gets energy density of space at the point T which is at the distance d from the centre. We can imagine that the radius of the stellar object is extended for distance d to the imaginary point T*and its volume is increased respectively. Because of the increased imaginary volume of the stellar object, the value of the energy density of space in its centre becomes bigger and is increases in the direction away from the centre. By using this imaginary volume we will actually calculate the energy density of space at the real point T. When $d=r$, one gets energy density of space on the surface of the stellar object. When $d=\infty$, one gets energy density of space in intergalactic space far away from stellar object which is Planck energy density ρ_{PE}.

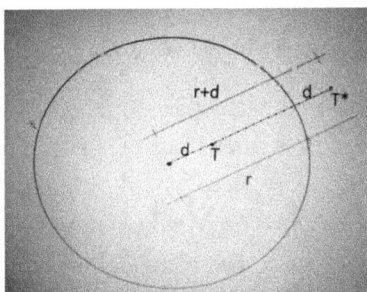

Figure 3: Energy density of space of a given point T from its centre

The calculation of the energy density of space on the Earth's surface is as follows:

$$\rho_{SE} = 4.633 \times 10^{113} J/m^3 - \frac{3(5.97219 \times 10^{24} kg)(299792458 ms^{-1})^2}{4\pi (2 \times 6371000 m)^3}$$

$$\rho_{SE} = 4.633 \times 10^{113} J/m^3 - \frac{(17.91657 \times 10^{24} kg)(8.99 \times 10^{16} m^2 s^{-2})}{4\pi (2.07 \times 10^{21} m)^3}$$

$$\rho_{SE} = 4.633 \times 10^{113} J/m^3 - \frac{1.6107 \times 10^{42} m^2 s^{-2}}{2.6012 \times 10^{22} m^3}$$

$$\rho_{SE} = 4.633 \times 10^{113} J/m^3 - 6.192 \times 10^{19} J/m^3$$

$$\Delta_{space.energy.density} = \rho_{PE} - \rho_{SE} = 6.192 \times 10^{19} J/m^3.$$

The calculation of the energy density of space at the Earth's centre is as follows:

$$\rho_{SE} = 4.633 \times 10^{113} J/m^3 - \frac{3(5.97219 \times 10^{24} kg)(299792458 ms^{-1})^2}{4\pi (6371000 m)^3}$$

$$\rho_{SE} = 4.633 \times 10^{113} J/m^3 - \frac{(17.91657 \times 10^{24} kg)(8.99 \times 10^{16} m^2 s^{-2})}{4\pi (2.586 \times 10^{20} m)^3}$$

$$\rho_{SE} = 4.633 \times 10^{113} J/m^3 - \frac{1.6107 \times 10^{42} m^2 s^{-2}}{(3.250 \times 10^{21} m)^3}$$

$$\rho_{SE} = 4.633 \times 10^{113} J/m^3 - 4.956 \times 10^{20} J/m^3$$

We see here that the space energy density on the Earth's surface is smaller than the space energy density in outer space (which is Planck energy density $\rho_{PE} = 4{,}633 \cdot 10^{113} J/m^3$) by a value of $6{,}192 \cdot 10^{19} J/m^3$. On the Earth's surface in the centre of a proton vortex the space energy density is smaller by $6{,}015 \cdot 10^{-66} J/m^3$. Energy density of space in the centre of the Earth is $4{,}956 \cdot 10^{20} J/m^3$ smaller than Planck energy density. In the centre of black hole with the mass of the sun and radius of 3000 m energy density of space is smaller than Planck energy density for $1{,}582 \cdot 10^{36} \, J/m^3$.

Planck energy density P_{PE}	$P_{PE} = 4{,}6 \times 10E113 \, J/m3$
Energy densiy on the Earth surface	$P_{PE} - 6{,}6 \times 10E19 \, J/m3$
Energy density in the Earth centre	$P_{PE} - 4{,}9 \times 10E20 \, J/m3$
Energy density in the centre of the proton	$P_{PE} - 6{,}0 \times 10E34 \, J/m3$
Energy density in the centre of black hole	$P_{PE} - 1{,}6 \times 10E36 \, J/m3$

Figure 4: Calculations of space energy density

In figure 4 we can see an extremely interesting thing; namely, the energy density at the centre of a proton is smaller than at the centre of the Earth. This gives a new interpretation of the "strong force" which binds together protons and neutrons. At the range of $10^{-15} m$ (1 femtometer), the strong force is approximately 137 times as strong as electromagnetism, a million times as strong as the weak interaction, and 10^{38} times as strong as gravitation. In the model proposed here, the strong force has its origin in the diminished energy density of space that exists at the centre of a proton's vortex. The gravitational force F_g between stellar objects and the strong force inside an atom's nucleus both have origin in the diminished energy density of space where protons and neutrons are pushed together in the same way as stellar objects via the pressure of the higher energy density of outer space towards lower the energy density of the space around the stellar objects, according to the famous Newton formula:

$$F_g = \frac{m_1 \cdot m_2 \cdot G}{r^2} \quad (18),$$

where m_1 is the amount of space energy incorporated in the first stellar object, m_2 is the amount of space energy incorporated in the second stellar object, G is the gravitational constant, and r is the distance between the centres of the stellar objects:

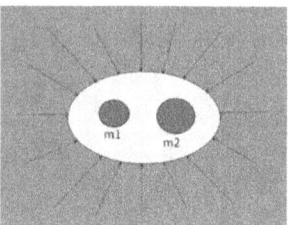

Figure 5: Gravity as the result of the higher outer pressure of space

The diminishing of the space energy density towards the centre of a given massive object is a missing model of today physics, which gives us a clear understanding about the relation: "massive object – space". The concept of the energy density of space in this article has bijective correspondence with the curvature of space utilized in General Relativity. Specifically, the greater the energy density of space the less is its curvature, whereas the less dense space is the greater is its curvature (Fiscaletti, Sorli 2015; Fiscaletti, Sorli 2017).

Neutron when alone is not stable and is not stable vortex as proton, electron and photon. Neutron is not fundamental primordial vortex; it is composed of proton vortex and electron vortex. Photon is also vortex of space energy. Elementary vortexes with stable lifetimes (proton, electron and photon) we can name "real particles" in the sense that they exist on their own as the constitutive elements of the universe. Particles with short lifetimes are only characteristic fluxes of energy which are released in a given physical process.

For example, the lifetime of pions (or "pi mesons", denoted with Greek letter π) is $2{,}6033 \cdot 10^{-8} s$. We should not consider that pions are particles in the same way that protons, electrons and photons are considered to be particles.

SUSY, which predicts the existence of supersymmetry where every particle has its partner particle, is as we have seen in chapter 5, an incorrect model. The supersymmetry in the universe is not between particles, it is between the mass of the proton and the diminished energy density of space in the centre of the proton vortex. This symmetry is universal from the micro to the macro scale; it gives origin to the inertial mass of the proton, the inertial mass of an atom's nucleon, and the inertial mass of a given physical object or stellar object.

One could ask: How it is possible that space has such a high energy density and we cannot detect it? The answer is as follows: the energy of space is the primordial energy of the universe, and syntropy is one of its characteristics (Sorli et al., 2016; Sorli et al., 2017b). Protons, electrons, and photons are vortexes made of this syntropic energy, which is what allows them to have such an extended lifetime. Entropy starts with the atoms which, in the course of universal dynamics, have the characteristic of increasing entropy as we progress up the ladder from helium to the atoms with big atomic numbers, and which atoms in black holes finally disintegrate back into the primordial syntropic energy of space (Sorli et. all., 2017b). Electrons do not fall into the nucleus of atoms because electrons are vortexes of space energy, and so are syntropic. Electrons do not get "exhausted" after spinning billions of years around a nucleus because electrons do not need (in the sense of classical mechanics) energy for their motion. In general we can say that universe is the "perpetual mobile" and that the energy of space is an inexhaustible source of energy. With the introduction of space as the fundamental syntropic energy of the universe, the theoretical foundations for "free energy technology" and antigravity have been laid. Once we are able to develop technology that makes it possible to increase and diminish the energy density of space, we will then be able to develop antigravity vehicles.

In this article it has been shown that the vortex model of elementary particles has a stronger bijective correspondence with the physical world than the Standard model, because the vortex model clearly describes the relation between mass (as the amount of energy incorporated in a given particle), inertial mass, and gravitational mass. Other researchers have proposed a vortex model of elementary particles where the mass and charge are global and indivisible properties of vortices (Rockenbeur, 2009). The vortex particle model presented in this article follows Ervin Schrödinger's vision: *"What we observe as material bodies and forces are nothing but shapes and variations in the structure of space"* (Sorli et al., 2018). The 20th century idea of physics that universal space has only geometrical properties, and is without physical properties, has lead modern physics into a "Standard model crisis," where more and more new particles are discovered and yet still the unification of gravity with Standard model particle physics seems far away. The vortex model of the proton with variable energy density of space is a novel model in the unification of gravity with particle physics.

8. Conclusions

Bijective analysis is an indispensable tool for the validation of physical equations and physical models. For mathematical physics, successful testing with bijective analysis provides an assurance that a given physical equation or a given physical model is a 100% exact picture of the observed physical reality. In the absence of fully respecting bijective analysis, it is quite possible that described phenomenon exists only in the mind of the scientist without having the actual physical existence.

References:

Abbott B.P. et al., Observation of Gravitational Waves from a Binary Black Hole Merger, Phys. Rev. Lett. 116, 061102 – Published 11 February 2016
http://dx.doi.org/10.1103/PhysRevLett.116.061102

Bezrukov, F. & Shaposhnikov, M. Higgs inflation at the critical point. *Physics Letters B* 734, 249–254 (2014)

CERN,Supersymmetry,(2018)
https://home.cern/about/physics/supersymmetry

Dirk K.F. Meijer, Hans J.H. Geesink, Consciousness in the Universe is Scale Invariant and Implies an Event Horizon of the Human Brain, NeuroQuantology Volume 15, Issue 3, Page 41-79 (2017)

Dirk K. F. Meijer and Hans J. H. Geesink, Phonon Guided Biology: Architecture of Life and Conscious Perception Are Mediated by Toroidal Coupling of Phonon, Photon and Electron Information Fluxes at Discrete Eigenfrequencies, NeuroQuantology , Volume 14, Issue 4, Page 718-755 (2016)

Einstein A, Does the inertia of a body depends on its energy- content?, Annalen der Physik, 17, 1905

Fiscaletti, D. & Sorli, A. Bijective Epistemology and Space–Time, Found Sci (2015) 20: 387. https://doi.org/10.1007/s10699-014-9381-z

Fiscaletti D. and Sorli, A.: "Space-time curvature of general relativity and energy density of a three-dimensional quantum vacuum", Annales UMCS Sectio AAA: Physics 69, 55-81 (2014).

Fiscaletti D. Sorli A. Space-time curvature of general relativity and energy density of a three-dimensional quantum vacuum. Annales UMCS Sectio AAA: Physics 2015;69(1): 53-81

Fiscaletti D. Sorli A. Quantum vacuum energy density and unifying perspectives between gravity and quantum behaviour of matter, Annales de la Fondation Louis de Broglie, Volume 42, numéro 2, 2017

Gesicki K., Zijlstra A.A., Bertolami M., The mysterious age invariance of the planetary nebula luminosity function bright cut-off, Nature Astronomy, 7 May (2018)

Haramein N., Quantum Gravity and the Holographic Mass, Physical Review & Research International 3(4): 270-292, (2013)

NASA https://map.gsfc.nasa.gov/universe/uni_shape.html (2014)

Oldershaw R.L., Hadrons As Kerr-Newman Black Holes, https://arxiv.org/abs/astro-ph/0701006 (2010)

Piskounova Olga, Rapidity Gaps in Double Diffraction Events at LHC as a Manifestation of String Junction Net on the Topology of Torus, https://arxiv.org/pdf/1702.02769.pdf (2017)

Real-World Relativity: The GPS Navigation System (2017) http://www.astronomy.ohio-state.edu/~pogge/Ast162/Unit5/gps.html

Rezzolla L., Most E.R., Weih L.R. Using Gravitational-wave Observations and Quasi-universal Relations to Constrain the Maximum Mass of Neutron Stars, The Astrophysical Journal Letters, Vol. 852, Num.2 (2018)

Rockenbauer A., Can the spinning of elementary particles produce the rest energy m c2 ? The vortex model of elementary particles https://arxiv.org/ftp/arxiv/papers/0808/0808.1656.pdf (2009)

Sorli A., Fiscaletti D., Mageshwaran M., Unification of Space, Matter and Consciousness, NeuroQuantology, Vol.14, Num.4 (2016)

Sorli A., Dobnikar U., Fiscaletti D., Kaufman S., Advanced Relativity: Multidimensionality of Consciousness and Mind, Origin of Life, Psi Phenomena, NeuroQuantology, Vol. 15, Num. 2, Page 109-117, (2017a)

Sorli A., Kaufman S., Dobnikar U., Fiscaletti D. Advanced Relativity for the Renaissance of Cosmology and Evolution of Life, NeuroQuantology, Vol. 15, Vol. 4, Page 37-44, (2017b)

Sorli A., Kaufman S., The Epistemological Crysis in Modern Physics, NeuroQuantology, Vol. 16, Vol. 2, Page 1-5 (2018)

Sorli A., Dobnikar U., Patro S.K , Mageshwaran M., Fiscaletti D., Euclidean-Planck Metrics of Space, Particle Physics and Cosmology, NeuroQuantology, Vol. 16, Num. 4 , Page 18-25, (2018)

Tiec A., Novak J., Theory of Gravitational Waves, (2017) https://arxiv.org/pdf/1607.04202.pdf

Euclidean-Planck metrics of space, particle physics and cosmology

Abstract

Recent research has introduced a novel model where a fundamental arena of the universe is infinite Euclidean space of Planck metrics, where time is merely mathematical parameter of universal changes. The history of the universe has merely a mathematical existence and is nonexistent in the physical sense. On the other hand, the future is not yet existent. The only existent physical reality is the universe, which exists in the timeless space of Euclidean-Planck metrics. This view is the basis of an "Energy-Mass-gravity" Model that unifies energy, mass, and gravity. Additionally, this model reveals some discrepancies in the Big Bang cosmology model that need to be examined in details in order to keep the Big Bang cosmology as the leading model of today's physics.

Key words: Euclidean space, Planck units, energy, mass, gravity, cosmology, gravitational constant G.

1. Introduction

Recent research confirms that material changes run in timeless space. The linear time of "past-present-future" belongs to the mind. The time we measure with clocks is the duration of material changes, i.e. motion in timeless space (Sorli et al., 2017)

In timeless space every physical object and every signal moves in space only, and not in time. This related understanding has far-reaching implications for the field of astronomy and cosmology. For example, although it might take a few billion light years for a signal from a distant star to arrive at the Earth, in which case the star has already died, we nonetheless have to understand that the star has both emitted the signal, and died, in the same timeless space.

NASA results confirm that universal space has the form of Euclidean space, which is infinite: "We now know (as of 2013) that the universe is flat with only a 0.4% margin of error. This suggests that the Universe is infinite in extent; however, since the Universe has a finite age, we can only observe a finite volume of the Universe. All we can truly conclude is that the Universe is much larger than the volume we can directly observe" (NASA, 2013).

2. "Euclidean-Planck metrics" (EPM) of universal space

The Euclidean-Planck metrics of universal space are developed from the standpoint of considering Universal space as being timeless and as having a Euclidean shape. In set theory a set A is a subset of a set B, or equivalently B is a superset of A, if A is "contained" inside B, that is, all elements of A are also elements of B.

$$A \subseteq B \quad (1),$$
$$B \supseteq A \quad (2).$$

That A is the subset of B is denoted with (1). That B is the superset of A is denoted with (2).

We observe that in universal space there exists matter and electromagnetic energy, while theoretical research predicts the existence of dark matter and dark energy. Universal space S_U has properties of superset B (in sense that all elements of A are also elements of B).

Matter M, electromagnetic energy EM, dark matter M_D and dark energy E_D have properties of subset A. We can write this in the following form:

$$S_U : \{M, EM, M_D, E_D\} \quad (3).$$

Out of (3) it follows that the set universal space S_U must also have physical properties, as it has as its subsets M, EM, M_D, E_D. The idea of 20th century physics of an empty space deprived of physical properties that contain matter and energy does not seem to be correct. By taking into account NASA's discovery that universal space has a Euclidean shape, combined with the idea that Planck units represent possible physical properties of universal space, we are able to develop the "Euclidean-Planck metrics" of universal space (EPM) in which each Planck volume l_P^3 of space contains an amount of Planck energy E_P, which means that empty universal space devoid of matter and fields has a Planck energy density ρ_{PE}:

$$\rho_{PE} = \frac{E_P}{l_P^3} \quad (4).$$

Figure 1: Euclidean-Planck metrics (EPM) of universal space

Every elementary particle with an amount of energy E and without inertial mass (as for example a photon) will change the Euclidean-Planck metrics of universal space (further on EPM) in the sense that it will diminish the Planck energy density of space by exactly the amount of its energy E:

$$\rho_{SE} = \rho_{PE} - \frac{E}{l_P^3} \quad (5).$$

$$E = (\rho_{PE} - \rho_{SE}) \cdot l_P^e \quad (6),$$

Where $(\rho_{PE} - \rho_{SE}) = \Delta_{EPM}$, so we can write:

$$E = \Delta_{EPM} \cdot l_P^3 \quad (7).$$

Every massive particle will change the Euclidean-Planck metrics, meaning every massive particle will diminish the Planck energy density of space ρ_{PE} in its centre as follows:

$$\rho_{SE} = \rho_{PE} - \frac{mc^2}{V} \quad (8),$$

Where m is its mass and V is its volume.
Out of equation (8) we derive the equation for mass-energy equivalence as follows:

$$E = mc^2 = \Delta_{EPM} \cdot V \quad (9).$$

A relativistic particle, because of its high speed, creates a "dragging effect" within universal space. Owing to this dragging effect, the energy of space is additionally absorbed by the relativistic particle, and that is why a relativistic particle gains its relativistic energy, which can be expressed by the formula:

$$E = \gamma \cdot \Delta_{EPM} \cdot V \quad (10),$$

where γ is the Lorentz factor.

Formula (9) is also valid for massive objects and stellar objects. For massive objects with a given velocity v and volume V the formula for its total energy is as follows:

$$E = \Delta_{EPM} \cdot V + \frac{\Delta_{EPM} \cdot V \cdot v^2}{2c^2} \quad (11).$$

Formula (11) we can develop:

$$E = \Delta_{EPM} \cdot V \cdot \left(1 + \frac{v^2}{c^2}\right) \quad (12).$$

Formula (12) shows that the kinetic energy of a moving body has its origin in a diminished value of EPM. The kinetic energy of a moving body is the energy of space that is additionally stored in that moving body. If that body hits a wall, its kinetic energy will be released as heat and light. In LHC part of the kinetic energy of two protons colliding is released as a Higgs boson.

Out of equation (10) we get the relation between the relativistic energy of a particle, the diminished value of EPM, and the Lorentz factor:

$$\gamma = \frac{E}{\Delta_{EPM} \cdot V} \quad (13).$$

Now, one can write the following formula for the relativistic rate of clocks in SR

$$\Delta t_0 = \Delta t \cdot \frac{mc^2}{\Delta_{EPM} \cdot V} \quad (14),$$

where Δt_0 is the elapsed time in a moving inertial system (in case of GPS satellite) and Δt is the elapsed time in the stationary inertial system (in case of GPS surface of the Earth) and m is the mass of the satellite. The Lorentz factor is primarily related to the diminished energy density of space caused by the dragging effect between space and the satellite. Δt_0 and Δt depend exclusively on the variability of the EPM, and not on some special position of the observer in the sense of an "inner observer" and an "outer observer." GPS system proves this beyond any doubt (Ashby, 2003).

3. Euclidean-Planck metrics and "Energy-Mass-Gravity" Model

Albert Einstein used to say: *"The mass of a body is a measure of its energy-content"* (Einstein 1905). In other words, according to Einstein, mass and energy are made out of the same "stuff" and can be converted into each other. According to Ervin Schrödinger, space is the fundamental energy of the universe: *"What we observe as material bodies and forces are nothing but shapes and variations in the structure of space"* (Huntley 2013). In this article, a combined Einstein/Schrödinger view is developed by way of an "Energy-Mass-Gravity Model" (EMG Model) that will be presented in this chapter. In EMG Model universal space has properties of Bose-Einstein condensate and is in symmetry with all particles: every particle changes EPM of space in the sense that diminishes its energy density with respect to the Planck energy density exactly for the value of its energy and so mass. In EMG curvature of space-time, from the micro to the macro scale, represents only the mathematical description of the energy density of space. The changes of the energy density with respect to the EPM of space generate a curvature of space-time similar to the curvature produced by a "dark energy" density (Fiscaletti and Sorli, 2014; Fiscaletti, 2016), through a quantized metric, characterizing the underlying microscopic geometry of space, expressed by relation

$$d\hat{s}^2 = \hat{g}_{\mu\nu} dx^\mu dx^\nu \quad (15).$$

In equation (15) the (quantum operators) coefficients of the metric are defined (in polar coordinates) as

$$\hat{g}_{00} = -1 + \hat{h}_{00}, \quad \hat{g}_{11} = 1 + \hat{h}_{11}, \quad \hat{g}_{22} = r^2(1 + \hat{h}_{22}), \quad \hat{g}_{33} = r^2 \sin^2\vartheta(1 + \hat{h}_{33}),$$
$$\hat{g}_{\mu\nu} = \hat{h}_{\mu\nu} \text{ for } \mu \neq \nu \quad (16)$$

and

$$\langle \hat{h}_{\mu\nu} \rangle = 0 \quad \text{except} \quad \langle \hat{h}_{00} \rangle = \frac{8\pi G}{3}\left(\frac{\Delta\rho_{EPM}}{c^2} + \frac{35Gc^2}{2\pi\hbar^4 V}\left(\frac{V}{c^2} \Delta\rho_{EPM}^{DE} \right)^6 \right) r^2$$

$$\text{and} \quad \langle \hat{h}_{11} \rangle = \frac{8\pi G}{3}\left(-\frac{\Delta\rho_{EPM}}{2c^2} + \frac{35Gc^2}{2\pi\hbar^4 V}\left(\frac{V}{c^2} \Delta\rho_{EPM}^{DE} \right)^6 \right) r^2 \quad (17).$$

In this scheme, dark energy is itself structured energy of space on the basis of equation

$$\rho_{DE} \cong \frac{35Gc^2}{2\pi\hbar^4 V}\left(\frac{V}{c^2} \Delta\rho_{EPM}^{DE} \right)^6 \quad (18).$$

This means, taking account the results of Santos (2009, 2010) about the link between the two-point correlation function of the vacuum fluctuations and the space-time curvature, that the variable energy density corresponding to the dark energy acts as a two-point correlation function according to relation

$$\frac{c^4}{4\pi\hbar^4}\left(\frac{V}{c^2} \Delta\rho_{qvE}^{DE} \right)^6 \cong \int_0^\infty C(s) s \, ds \quad (19)$$

where $C(s)$ is the two-point correlation function of the fluctuations with respect to the value of the Planck energy density of EPM of space, which depends only on the distance between the two points. In EMG model, in the light of equations (15)-(19), the three-dimensional space defined by the quantized metric (15) determined by the changes and fluctuations with respect to the Planck energy density of EPM of space can be considered as the fundamental origin of the curvature of space-time characteristic of general relativity. In other words, there is a fundamental physical equivalence between curvature of space and diminishing of the energy density of the space.

According to equations (15)-(19), each form of energy has the property to modify the EPM of space, by generating the curvature of the space-time characteristic of general relativity. This means, in the light of equations (5)-(7), that in the EMG approach also the energy of a photon can cause a curvature. The physical origin of the curvature of space (and thus of the modification of the EPM of space) in the lowering of energy with respect to the Planck energy, implies therefore that also a photon, which has energy, as a consequence has got a mass in the sense of "mass as the amount of energy." A photon's mass can be defined by the following formula:

$$m = \frac{h \cdot v}{c^2} \quad (20),$$

where h is Planck's constant, v is frequency, and c is light speed. Unlike a proton, a photon has no rest mass, but its energy can also be presented as mass according to the mass-energy equivalence principle as well as the physical origin of the curvature of space in each form of change of energy density with respect to the Planck energy density. We could say that formula (20) shows that the photon's energy is equivalent to its "kinetic mass".

A given massive particle that is moving with velocity v mass m is the sum of rest mass m_0 and kinetic mass m_K:

$$m = m_0 + m_K \quad (21),$$

where m_K is:

$$m_K = \frac{m_0 v^2}{2c^2} \quad (22).$$

Moving particles interact with space in a so called "dragging effect" which increases it energy and mass. Kinetic mass m_K is the energy of space which is additionally integrated in the moving particle.

Combining formula (21) and (22) we get:

$$m = m_0 + \frac{m_0 v^2}{2c^2} \quad (23)$$

and thus

$$m = m_0 \cdot \left(1 + \frac{v^2}{2c^2}\right) \quad (24).$$

Also gluons, which represent 99% of proton mass, can be imagined as particles that have no rest mass, they only have kinetic mass that is equal to their energy. The same goes for quarks; i.e., we can imagine them as particles with no rest mass; they have their energy and correspondent kinetic mass. The rest mass of a proton m_0 is the sum of the kinetic masses of gluons $m_{K.Guons}$ and kinetic masses of quarks $m_{K.Quarks}$:

$$m_0 = \sum m_{K.Quarks} + \sum m_{K.Gluons} \quad (25).$$

Formula (25) can also be written as follows:

$$m_0 = \sum \frac{E_{Quarks}}{c^2} + \sum \frac{E_{Gluons}}{c^2} \quad (26).$$

In formula (26) the energy of quarks is presented according to the Schrödinger view (Huntley 2013), which means that it corresponds to the structured energy of space, and which view is also our view. This view does not need the existence of some special field (Higgs field) in order to give quarks mass. The Higgs mechanism represents the ultimate in complexity in particle physics while not contributing to the clarity of physics. More than that: the Higgs mechanism has created a gap between mass and energy that is contrary to their unification as represented in Einstein's "mass-energy" equivalence principle. Adding to this the epistemological instability of the Higgs mechanism (Sorli, Kaufman, 2018), we assume that the Higgs mechanism will not have a long "life-time."

Particles that have rest mass are different structures of space energy, and they diminish EPM exactly for the amount of their mass-energy according to formula (9). A diminished EPM, and therefore the corresponding equivalence between curvature of space and each lowering of the energy density with respect to the Planck energy density, is the physical origin of both inertial mass and gravitational mass. The pressure of outer space, which has a relatively higher EPM, moving towards the centre of a massive particle, which has a relatively diminished EPM, is what gives birth to both inertia and gravity.

In this way, two massive particles or physical objects create an area of diminished EPM that is the origin of gravity. In this model, outer space has a relatively higher EPM than the space immediately surrounding two physical objects. This creates an energy gradient or pressure in the direction of the particles that produce the diminished EPM. In essence, the pressure of outer space pushes together inner space that has a diminished EPM as a result of the presence of the two physical objects. In this way, particles and physical objects that are in space are pushed together indirectly via space. In this model, gravity does not work directly between two massive bodies. Rather, in this model, gravity works on bodies indirectly via the energetic structure of space.

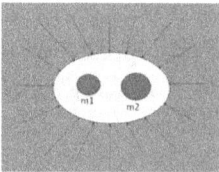

Figure 2: Gravity as the pressure of outer space that is created by the diminished EPM around two particles

The model presented in this chapter regarding the origin of energy, mass, and gravity (EMG Model) works both without a hypothetical graviton, as well as without a Higgs field. The origin of the energy and mass of all massive particles is a diminished EPM. The Higgs mechanism is developed upon the proposition that, in general, all particles are mass-less. Particles which interact with the Higgs field (for example quarks) will gain mass, while particles which do not interact with the Higgs field (for example photons) will not gain mass. The Higgs mechanism does not continue the tradition of Einstein's view, in which mass and energy are made out of the same "stuff". The Higgs model has several epistemological instabilities that need to be carefully examined (Sorli A., Kaufman S.,2018).

The Higgs mechanism or model has its origin in supersymmetry, which is an extension of the Standard Model: "Supersymmetry predicts a partner particle for each particle in the standard model, to help explain why particles have mass. At first sight, the Standard Model seems to predict that all particles should be massless, an idea at odds with what we observe around us. Theorists have come up with a mechanism to give particles masses that requires the existence of a new particle, the Higgs boson" (CERN, 2018). In the EMG Model the fundamental symmetry of the universe is between a given massive particle (or physical object) and the variable EPM, which was expressed in formula (9). In the EMG Model the Higgs boson corresponds mathematically to a flux of released relativistic energy caused by the collision of two protons and its action physically derives from a more fundamental interplay of opportune fluctuations of the energy density with respect to the EPM of space. The manmade artificial flux generated in the collision of two protons has an extremely short life-time of $1,56 \cdot 10^{-22} s$ and does not prove the existence of the hypothetical Higgs field, inasmuch as it is indirect evidence, and not the direct evidence that is, or at least should be, required to establish proof (Sorli, Kaufman, 2018). The main theoretical failure of Higgs mechanism is the prediction that some field is giving mass to the particles without considering that if so this field should give to the particles also energy, because energy and mass are made out of the same stuff, and there is a physical equivalence between curvature of space and each form of diminishing of the energy density with respect to the Planck energy density. In EMG Model there is no difference between mass and energy. Space is the source of mass and energy of all particles.

In the EMG Model, energy, mass, and gravity are intrinsically related to the variable EPM; the kinetic energy of a massive body additionally diminishes the EPM, thereby increasing the gravity force. Let's do a "thought experiment": we place two iron balls on a vertical axis at distance d and we then measure the gravitational force. Then we start rotating the balls, thereby giving them a high angular velocity, and we measure their gravitational force, and find that it is greater than that of the first measurement:

$$F_{g.rotating} \succ F_{g.still} \quad (27).$$

This thought experiment has theoretical support in previous research regarding the relativistic mass of a rotating cylinder (Gilloch J.M., W.H. McCrea, 1951).

4. Euclidean-Planck metrics and CMB signal

Euclidean-Planck metrics introduces the idea that CMB moves in timeless space and that time is the duration of its motion. CMB radiation has its source in the period of recombination, circa 377,000 years after the Big Bang. We can imagine this epoch as a slice of a three dimensional ball (shown in figure 3 below marked as **RS**). At the time of recombination, and so when the universe was around 377,000 years old, the epoch radius of the universe was around 42 million light-years. Because recombination lasted around 100,000 years, the source of CMB lasted around 100,000 years, and so ended when the age of the universe was around 477,000 years. This means that the source of CMB has not been physically present in the universe for 13.7 billion years minus 477,000 years. Therefore, CMB is relic radiation of a source that was "extinguished" around 13,699523 billion years ago. As shown in figure 3 below, given that CMB radiation was produced in the recombination epoch, the signal is now reaching an area in universal space that is 13,699523 billion light years distant from the radius of the recombination epoch (point **A** in figure 3). Given as well that the radius of the universe today is around 46,6 billion light years, this means that CMB should not be reaching us yet and is around 32,9 billion light years distant from the planet Earth (point **B** on the figure 3). This discrepancy needs to be solved in order to for CMB to continue as the main proof underlying Big Bang cosmology.

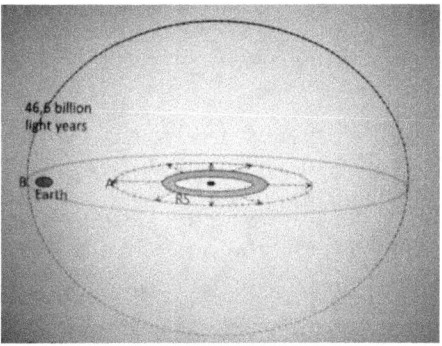

Figure 3: Plane of intersection of expanding universe

NASA results provide strong evidence that universal space has the shape of Euclidean space, which excludes the use of non Euclidean geometries in cosmology. Universal space does not have an open curvature, it does not have a closed curvature, universal space is flat. We can imagine the inflation of the universe as a balloon that is inflating in an infinite Euclidean space. Imagine that the idea that it is the expansion of the universe that is creating universal space is wrong. And then imagine instead that galaxies are moving away from each other in a stationary infinite space. The common understanding today, which is that the distances between galaxies are increasing because universal space is thought to be expanding like an inflating balloon, is not correct with respect to the results of NASA, because the NASA results show that "the Universe is infinite in extent," and that which is infinite cannot expand.

In the Big Bang model the estimated age of the universe is 13.7 billion years. The radius of the observed mapped universe is 46,6 billion light years. This means that, according to the Big Bang model, the universe has been expanding since its beginning at 3 - 4 times the speed of light, (1.020.437 kilometres per second) which is against both existent physical laws and common sense.

The velocity of accelerated expansion today is valued at $68 kms^{-1}$ which puts (considering the age of the universe is 13,7 billion years) the radius of the universe at $2,78 \cdot 10^{19} km$. On the other hand, the radius of the observed universe is around $4,41 \cdot 10^{23} km$. This discrepancy of the rate 10^4 indicates there is a huge system error in the Big Bang model regarding the size of the observable universe.

5. Universe in dynamic equilibrium has stable value of gravitational constant G

We propose in this article a model of the universe in dynamic equilibrium. In intergalactic space, where the value of EPM is at a maximum, the energy of space is continuously transforming into cosmic rays that then themselves transform into elementary particles (Friedlander, 2002). In the centre of black holes the EPM value is at a minimum, and as a result atoms become unstable and disintegrate back into the energy of space. The calculation of EPM in the centre of black hole that has a mass M that is equal to the mass of the Sun, and has a radius r of 3000 metres gives the following value:

$$\rho_{SE} = \rho_{PE} - \frac{3M \cdot c^2}{4\pi r^3} = 4{,}633 \cdot 10^{113} J/m^3 - 1{,}582 \cdot 10^{36} J/m^3$$

(28)

and thus here one has

$$\Delta_{EPM} = 1{,}582 \cdot 10^{36} J/m^3 \quad (29).$$

The value of EPM in the centre of a black hole the size of the Sun is smaller than in outer intergalactic space by $1{,}582 \cdot 10^{36} J/m^3$.

The circulation of energy just described, i.e., the process "formation of particles in outer space - formation of stars - black holes - disintegration of matter in space energy," is eternal; it has no beginning and will not have an end. Black holes are the "rejuvenating mechanisms" of the universe, where "old" matter is transformed back into the "fresh" energy of space itself. The universe as a whole has an infinite amount of energy which cannot be created and cannot be destroyed.

In a universe of dynamic equilibrium, the gravitational constant G is unchangeable. The value of the gravitational constant G can be written as follows:

$$G = \frac{l_P^3}{m_P \cdot t_P^2} = \frac{1}{\rho_P \cdot t_P^2} = \frac{c^2}{\rho_{PE} \cdot t_P^2} \quad (18).$$

The calculation of G in the centre of a black hole, where EPM diminishes by $1{,}582 \cdot 10^{36} J/m^3$, confirms that the value of G remains unchangeable in the measurable rate and we can then consider that it is of the same value throughout universal space (Sorli at al., 2018). Our research group plans an experiment to measure the value of G at three different places on the globe (India, Russia, China) at the same time periodically every month for one year. In this way we will get statistically significant data about G values.

In the model that is being presented, which is a model of the universe in a permanent dynamic equilibrium, dark energy is the energy of space itself (Fiscaletti 2016). Space is neither empty nor filled with some type of energy; rather, space is the concrete fundamental energy of the universe as was proposed by Ervin Schrödinger (Huntley 2013). The curvature of space in General Relativity is the mathematical expression of its energy density. A higher curvature of space means a lower energy density of space (Fiscaletti, Sorli 2015) for which the actual geometry is Euclidean.

6. Cosmology without paradoxes

At night, we see the universe as a dark sky simply because the light coming from the galaxies is not strong enough to lighten all of universal space. When you put candles in a very large room, the room remains dark and you see the sources of light. You can imagine the universe as a room that is infinite in extension and has an infinite number of candles which are separated by enormous distances. The idea that universal space would be full of light if the universe was not expanding makes no sense. According to our model, in our observable space, the galaxies do not have enough light to fill all of universal space with light. Olber's paradox is a classic example of the way in which science sometimes creates a problem through the wrong reasoning.

The other problem with today's cosmology is the inflation model, which unsuccessfully tries to explain the appearance of energy in the first moments of the Big Bang. There is no reasonable explanation for the appearance of energy in the so-called "first moments" of the Big Bang. We have known in physics, since the time of Newton, that energy cannot be created and cannot be destroyed; but it can be transformed. The Big Bang model does not satisfy the law of conservation of energy. On the other hand, the model of the universe presented here, which describes the universe as being in dynamic equilibrium, fully satisfies the law of energy conservation.

The inflation signal could not be detected by BICEP2 (Cortês, 2015) because it could not have yet reached the Earth (as was also shown for the CMB signal, in chapter 4). In the cosmological model that we present in this article, the appearance of energy is not a problem. Another problem with Big Bang cosmology is in its failure to account for both where the energy comes from for the "the initial kick" for the explosion, as well as what exactly it is that has exploded. Our model has no such problems. Big Bang cosmology needs a "creator," a someone who has given an initial energy that causes the "birth" of the universe. In this sense, the Big Bang cosmology has some "biblical elements" that are not deserving of being a part of cosmology. In our model, the universe is a non-created system in a permanent dynamic equilibrium; it works perfectly without a creator.

7. **Conclusions**

The idea that particles and fields exist in an empty space deprived of physical properties is the main obstacle standing in the way of the progress of physics. With the introduction of Euclidean-Planck metrics (EPM) of universal space, particle physics gains a new model regarding the origin of energy, mass, and the gravity of elementary particles. More than this, EPM applied to cosmology shows that the Big Bang model has insufficiencies that are unsolvable. NASA results confirm that universal space is Euclidean, which excludes the possibility that universal space could be finite, which itself then excludes the possibility that it could expand. The model of infinite universal space governed by the EPM presented in this article heralds the end of the Big Bang model of cosmology and introduces a model of the universe that is in a permanent dynamic equilibrium.

References:

Ashby Neil, Relativity and the Global Positioning System, Living Rev Relativ. 2003; 6(1): 1.

CERN, Supersymmetry, https://home.cern/about/physics/supersymmetry (2018)

Einstein Albert, Does the inertia of a body depends on its energy- content?, Annalen der Physik, 17, 1905

Fiscaletti Davide, About Dark Energy and Dark Matter in a Three-Dimensional Quantum Vacuum Model, Found Phys (2016) 46: 1307.

Fiscaletti D. and Sorli, A.: "Space-time curvature of general relativity and energy density of a three-dimensional quantum vacuum", Annales UMCS Sectio AAA: Physics 69, 55-81 (2014).

Gilloch Josephine M. and W. H. McCrea, The relativistic mass of a rotating cylinder, Mathematical Proceedings of the Cambridge Philosophical Society, Volume 47, Issue 1, pp. 190-195. (1951).

Huntley Noel, The Emerging New Science, Author House, pp. 186. (2013)

NASA https://map.gsfc.nasa.gov/universe/uni_shape.html (2014)

Marina Cortês, Andrew R. Liddle, and David Parkinson, Tensors, BICEP2 results, prior dependence, and dust, Phys. Rev. D 92, 063511 (2015)

M. W. Friedlander, A Thin Cosmic Rain: Particles from Outer Space, Harvard University Press, Harvard. (2002).

Santos E., "Space-time curvature induced by quantum vacuum fluctuations as an alternative to dark energy", International Journal of Theoretical Physics **50**, 7, 2125-2133 (2010).

Santos E., "Dark energy as a curvature of space-time induced by quantum vacuum fluctuations", arXiv:1006.5543 (2010).

Sorli A., Kaufman S., Dobnikar U., Fiscaletti D., Advanced Relativity for the Renaissance of Cosmology and Evolution of Life, NeuroQuantology, December 2017, Volume 15, Issue 4, Page 37-44

Sorli Amrit, Santanu Kumar Patro, R.N.Patra, Relativistic Rate of Clocks and Stability of the Gravitational Constant G, International Journal for Research in Applied Science & Engineering Technology, Volume 6 Issue I, January 2018

Sorli A., Kaufman S., The Epistemological Crisis in Modern Physics, accepted in NeuroQuntology, February 2018

Minkowski space-time and Einstein's Now conundrum

Abstract

The Minkowski formula X4=ict indicates that time *t* does not express the 4th dimension of space-time, i.e., X4 is not *t*. Therefore, Minkowski space-time is not a 3D+T manifold, it is 4D. The 4D manifold has 4 spatial dimensions and is timeless, in the sense that it does not contain some physical time in which material changes run. In physics we experience the timelessness of 4D space as "Einstein's Now." From this perspective, the time that we measure with clocks is just the sequential numerical order of changes, i.e. motions that run in the timeless space of Now. On the other hand, past and future are nothing more than psychological or conceptual realities that derive from the neuronal activity of the brain. Therefore, "time flow" and "arrow of time" have only a mathematical existence.

Key words: Now, time, space, space-time, GPS, closed timeline curve, arrow of time.

1. **Introduction**

Today, 130 years after the pioneering work of Ernst Mach regarding the interpretation of time as a measure of the changes of things, the related ideas that time cannot be considered as a primary physical reality that flows on its own in the universe, and thus that universal space is timeless, are receiving more and more attention. For example, Girelli, Liberati and Sindoni, by following the philosophy of analog models of gravity, suggest that time and gravity might not be fundamental per se, but only emergent features, in the context of a toy-model where the Lorentzian signature and Nordström gravity (a diffeomorphisms invariant scalar gravity theory) emerge from a timeless non-dynamical space (Girelli et al., 2011). Moreover, as regards the meaning of time in a timeless background, interesting results have been obtained recently by Gózdz and Stefanska (Gózdz, Stefanska, 2008), Elze (Elze, 2002), Caticha (Caticha, 2011), Prati (Prati, 2011) and Anderson (Anderson, 2013). In particular, Prati showed that a physical system S, if complex enough, can be separated into a subsystem S2 whose dynamics is described, and another cyclic subsystem S1 which behaves as a clock and that, as a consequence of the gauge invariance, this separation may be made in many ways. Prati's model thus suggests, from the physical point of view, that the time provided by each subsystem that acts as a clock cannot be considered as an absolute quantity, indicating that the meaning of time in a timeless picture is as follows: the ticking of each subsystem acts as a clock that provides a different reference system describing the dynamics of that subsystem, and therefore that time, as an idealized quantity that flows on its own, does not exist.

An analysis of the Minkowski manifold confirms that space-time is timeless. In this way, "Einstein's Now" fully enters the realm of physics, in the sense that, in physics, we experience the timelessness of space as Now. In the Minkowski manifold the formula for the 4th dimension of space-time is as follows:

$$X4 = ict \ (1),$$

namely the fourth dimension is the product of the imaginary number i, light speed c, and time t. This formula has an internal structure that is similar to the formula for distance d, in which the imaginary number i is missing:

$$d = vt \quad (2),$$

where d is distance, v is speed, and t is time. The formula (1) confirms that the 4th dimension of space-time is not temporal, but is also spatial, as are the other three dimensions: X_1, X_2, X_3. Thus, the space-time manifold of Minkowski is not 3D + T. Therefore, the generalization that time is the 4th dimension of space is not appropriate. Rather, the space-time manifold of Minkowski is 4D. Minkowski's vision was to fully integrate space and time as an inseparable entity. On September 21, 1908, Hermann Minkowski began his talk at the 80th Assembly of German Natural Scientists and Physicians with the now famous introduction: "The views of space and time which I wish to lay before you have sprung from the soil of experimental physics, and therein lies their strength. They are radical. Henceforth space by itself, and time by itself, are doomed to fade away into mere shadows, and only a kind of union of the two will preserve an independent reality" (Minkowski, 1952). Minkowski was right. The mathematical formalism X4=ict confirms that, in his model, time is fully integrated within a timeless 4D space.

2. The solution to Einstein's Now conundrum

The research of Catalin V. Buhusi and Warren H. Meck confirms that the human experience of duration (time interval) is based on the neuronal activity of the brain (Catalin at al., 2005). Therefore, the linear time of "past-present-future" exists only as a function of the neuronal activity of the brain. We experience the run of changes in timeless space within the frame of the conceptual linear time of "past-present-future." Einstein himself was aware that the linear time of "past-present-future" was only a psychological reality. He used to say: "People like us, who believe in physics, know that the distinction between past, present and future is only a stubbornly persistent illusion" (Einstein 1915). The Minkowski manifold X4=ict, in which time t is just an internal element of the formula for the 4th imaginary dimension, is a mathematical model that is completely consistent with Einstein's view of time. In his book "Now: The Physics of Time" Richard Muller presents very directly Einstein's trouble with Now: "Albert Einstein was troubled with the concept of now." Further, philosopher Rudolf Carnap writes in his Intellectual Autobiography that Einstein said that the problem of the Now worried him seriously. Einstein explained that the experience of the Now means something special for man, something essentially different from the past and the future, but that this important difference does not and cannot occur within physics. That this experience cannot be grasped by science seemed to him a matter of painful but inevitable resignation. So Einstein concluded "that there is something essential about the Now which is just outside the realm of science" (Muller, 2016).

The solution with regard to bringing the Now into physics is through an understanding that material changes actually run in the physical reality of 4D timeless space, which we experience as "Now," and not in the psychological or conceptual reality of "past-present-future." In order to illustrate this, let us take an everyday example: walking from point A to point B we experience that our walking occurs in space, and always takes place Now. We do not perceive with our senses that we walk in time. We have a sensation of moving in time, because we experience our motion in the frame of the linear conceptual time of "past-present-future," which as previously stated, has only a psychological existence.

Another example of material changes actually running in the physical reality of 4D timeless space can be found by examining the motion of a photon in space. On the basis of a granular structure of space and time, which emerges from the results of loop quantum gravity, as well as other significant models, such as reticular space-time dynamics (Rovelli, 1998.; Rovelli 2011.;Licata 1991.). One can assume that quanta of space having the size of Planck length $l_p = \sqrt{\frac{\hbar G}{c^3}}$ are the fundamental constituents of space, and that Planck time $t_p = \sqrt{\frac{\hbar G}{c^5}}$ is the least unit of motion, and so represents the fundamental unit of numerical order of material changes (Fiscaletti, Sorli, 2015). For example, let us examine the motion of a photon in space. The photon passes Planck distance in Planck time, which is just the numerical order of the photon's motion. The photon is not moving in some physical time, as elementary perception and experimental data both confirm that the photon moves only in space. When we calculate the sum of the Planck times of the photon's motion from point A to point B we get time t as the duration:

$$t = t_{P1} + t_{P2} + ... + t_{PN} = \sum_{i=1}^{N} t_{Pi} \quad (3).$$

What this demonstrates is that changes run in timeless space, and that the existence of duration requires a measurement from the side of the observer. Equation (3) expresses the duration of the photon's motion from point A to point B, which is generated (and thus enters existence) as a consequence of the measurement of an observer. For this reason, we can call the time t that expresses duration the emergent time. Without a measurement there is no duration. Moreover, by taking into account that the existence of duration of physical events requires that the observer make a measurement, one can speculate that there are two ways to understand time:

– Time measured with clocks is a numerical order of change that has only a mathematical existence;
– Duration of a given material change requires that the observer makes a measurement.

These two ways to understand time indicate that in physics we have two kinds of time:

1. *Fundamental time*, which is the numerical order of change and exists independent of the observer.
2. *Emergent time*, which is the duration of material change and originates from the observer's measurement (Fiscaletti, Sorli, 2015).

An interesting possible mathematical characterization of fundamental time, namely of the numerical order of material changes, is represented by Prati's number k_{AB}, which provides a counter function of the number of states of the Hilbert space of the system of interest whose dynamics are being analysed, and that satisfies an appropriate initial condition (namely the origin of measurement) $\overline{\Psi}_1$ of the subsystem acting as a clock. By considering fundamental time as the numerical order of material changes defined by Prati's counter function of the states of a system k_{AB}, one obtains a suggestive unifying re-reading of the two fundamental theories of time existing in literature, namely the ephemeris time of the Jacobi-Barbour-Bertotti theory, and Rovelli's thermal time hypothesis. In our approach, Barbour's ephemeris time, the unphysical evolution parameter λ which provides a metric-trans-temporal notion of identity between two subsequent configurations, and thus a duration corresponding to the relative change in the positions of the particles in the system, can be considered as an emergent time that derives from the dynamics of the system, in the sense that it derives from the numerical order associated with the number k_{AB} of states of the Hilbert space of the system of interest whose dynamics are being described, on the basis of equation:

$$\int_{\lambda_0}^{\lambda_f} \frac{\sqrt{T}}{\sqrt{E-V}} d\lambda = k_{AB} \quad (4),$$

where V represents the potential energy of the system, and E the total energy of the system.

In an analogous way, Rovelli's "thermal time hypothesis" – according to which time is a reflex of our incomplete knowledge of the state of the world, i.e. a statement about the statistical state in which the world happens to be, when described in terms of the macroscopic parameters we have chosen – can be seen as a consequence of the fact that clocks measure merely the numerical order of material changes corresponding to Prati's number k_{AB} of states of the Hilbert space of the system of interest whose dynamics are being described. In other words, Rovelli's thermal time t_P (provided by the thermal clock whose ticking grows linearly with the Hamiltonian flow $s(t_P)$ of the nonrelativistic Hamiltonian of the system) can be considered as a statistical characterization of an emergent time which derives from the fundamental time represented by the numerical order of material changes defined by the counter function k_{AB} of states of the Hilbert space of the system of interest whose dynamics are being described, on the basis of equation:

$t_P = k_{AB}$ (5) (Fiscaletti, Sorli, 2015).

Einstein's famous quote confirms this view of time: "Time has no independent existence apart from the order of events by which we measure it" (Gomel, 2010). The view presented above solves the Now conundrum in some detail, as follows: Fundamental time is the sequential numerical order of changes that run in the Now, whereas emergent time is the duration of changes that run in the Now, and only comes into existence when measured by an observer.

Girelli, Liberati and Sindoni have shown in their article that time may not be fundamental, but may instead be an emerging feature: "We have showed, in a toy model, how the Lorentzian signature and a dynamical space-time can emerge from a flat non-dynamical Euclidean space, with no diffeomorphisms invariance built in. In this sense the toy-model provides an example where time (from the geometric perspective) is not fundamental, but simply an emerging feature" (Girelli et al., 2011). Our research confirms their results: time as duration is an emerging feature that only comes into existence when fundamental time is measured by an observer. Further, this way of conceiving of time provides a unifying re-reading of Barbour's ephemeris time and of Rovelli's thermal time, by including both of these ways of looking at time in a model where the concept of duration measured by an observer emerges from a more fundamental numerical order of material changes that exist independent of the observer.

3. **Discussion**

Andrew Jaffe is convinced that the Now is not a problem: "I'm convinced that 'now' is a non-problem. Once quantum mechanics and thermodynamics have given time a direction, 'now' isn't physics, but a combination of time's arrow with psychology and physiology. The past is what is encoded in our memories. To a rock, an electron or a galaxy, there is no now. But occasionally I wonder whether this is sufficient" (Jaffe, 2016). In physics we do not have any tangible experimental data that time has a physical direction, nor do we have any tangible experimental data that the physical arrow of time exists. Now is not a problem once one understands that time is the sequential numerical order of changes running in the timeless space of Now.

Shapiro has observed through measurement that in a relatively strong gravitational field light has a diminished speed. In an article entitled "Fourth Test of General Relativity" Shapiro wrote: "Because, according to the general theory, the speed of a light wave depends on the strength of the gravitational potential along its path, these time delays should thereby be increased by almost 2x10^{-4} sec when the radar pulses pass near the sun. Such a change, equivalent to 60 km in distance, could now be measured over the required path length to within about 5 to 10% with presently obtainable equipment" (Shapiro, 1964).

The solution to Einstein's Now conundrum presents us with a new physical model in which light moves in space, which is always Now, and time is its duration. In this new physical model, time does not exist as a physical reality (time as duration is only an emergent quantity which originates from observer's measurements) and as such, cannot shrink or dilate. The results of Shapiro confirm that in a relatively strong gravitational field the speed of light is diminished only minimally, because in a relatively strong gravitational field what actually changes are the permeability and permittivity of space, which changes only minimally influence light speed. In a relatively strong gravitational field light is moving a bit slower and so needs more duration to move from point A to point B than it does in space with weaker gravity. Understood in this way, it is more appropriate to refer to this effect as a "gravitational velocity decrease," rather than as a "gravitational time delay." Several pieces of research already confirm that gravity changes the permittivity and permeability of space, and that these changes diminish light speed (Sato, 2007., Ellis et al. 2007).

However, this effect applies not only to light speed, because the rate of all physical changes diminishes in a relatively stronger gravitational field. In GPS systems, clocks on the satellites are calculated to run 45 microseconds per day faster than clocks on the

Earth's surface, because gravity is stronger on the Earth's surface. This is the so-called "General Relativity (GR) relativistic effect." However, in GPS systems, clocks on the satellites are also calculated to run 7 microseconds per day slower than on the Earth's surface, owing to "Special Relativity (SR) time dilation," which would be better named "SR velocity decrease." Owing to this combination of GR and SR effects, the rate of clocks on the satellites is faster than the rate of clocks on the Earth's surface by 38 microseconds per day (Ashby, 2003).

However, the rate of clocks on the Earth's surface do not get slower because they exist in a physical reality called time that can actually dilate. In SR and GR (and in physics in general) time is only the mathematical parameter of material changes, i.e. motion, and cannot be "relative." We see in GPS systems that what is actually "relative" is the velocity of material changes i.e. motion. All clocks are running in the same timeless space of Now. In a relatively stronger gravitational field the rate of clocks is slower, the speed of light diminishes, and the velocity of physical and biological changes diminishes as well.

As regards the famous special relativistic argument of the twins, once it is realized that the twins are both aging in space, and not in time, there is no longer a "twin paradox." A twin on the Earth's surface ages faster than his or her counterpart on a fast spaceship owing to the SR relativistic effect. A Twin on the Moon ages faster than his or her counterpart on the Earth owing to the GR relativistic effect. The rate of the aging process for each twin is therefore different, but is real for both twins, because both twins actually only ever age in the timeless space of Now.

In a universe where the fundamental arena is timeless space, travelling into either the past or the future is categorically excluded. The idea of time travel, as something that might actually be possible, was born with the prediction of "closed time lines," or the more frequently used term, "closed timelike curve (CTC)," which were predicted by Willem Jacob van Stockum (Stockum, 1937) and Kurt Gödel (Gödel, 1949). However, Gödel acknowledged that CTC allows for contradictive time travel, leading him to conclude that time cannot have a physical existence. He expressed this view in a famous statement: "In any universe described by the theory of relativity, time cannot exist." (Yourgrau, 2006). Stockum and Gödel simply used the wrong term to describe the "closed time lines" that they discovered, and this in turn then led them to the erroneous conclusion that one could travel in time. In the first chapter of this article it was shown that the fourth coordinate of space is not temporal, but is also spatial. Hypothetical movement in "closed time lines" means that we move only in space, and so we always end up at the same point from which we started, which is in space Now. Motion in time is categorically excluded, because time is merely the duration of motion in *space*. The result of Hawking's research is that the laws of physics do not allow for the physical appearance of closed timelike curves in which one could move in time (Hawking, 1992).

Moreover, the idea of time as the sequential numerical order of changes running in the timeless space of Now allows us to justify, in a clear way, in what sense in general relativity the idea of an idealized time *t* that flows on its own in the universe, without reference to anything that happens, must be abandoned and be replaced with different possible internal times associated with specific physical clocks. Taking this approach allows one to deal with the relative motion of the variables, with respect to each other, in a democratic fashion. In general relativity, there is not a preferred and observable quantity that plays the role of independent parameter of the evolution of a system, because clocks provide only a mathematical measure of the numerical order of physical events in the timeless space of Now. With clocks one measures frequency, speed, and numerical order of events in the timeless space of Now. Since clocks can be defined as those instruments which measure the speed of material changes and movements, the internal clocks/times of general relativity are only measuring systems, and what they measure is the numerical order of material changes in the timeless space of Now. The definition of time as a mathematical coordinate that indicates the sequential numerical order of material motion running in the timeless space of Now thus provides a clear and suggestive re-reading of the nature, significance, and meaning of the internal clocks/times of general relativity (Fiscaletti at al., 2012)

Page and Wootters criticise modelling time as a physical dimension (coordinate time), because time as a physical dimension is unobservable: "We shall argue that the temporal behavior we observe is actually a dependence on some internal clock time, not on an external coordinate time. It is perfectly consistent with our observations to assume that any closed system is in an eigenstate of energy and thus stationary with respect to coordinate time, since coordinate time translations are unobservable. Such a state can be decomposed into states of definite clock time. The dependence of these component states upon the clock time labeling them can then represent the observed temporal behavior of the system" (Page, Wootters,1983). In this article it has been shown that coordinate time (temporal dimension) is also spatial. Clocks run in a timeless 4D space. The relativistic rate of clocks depends on the strength of the gravitational field in a given region of 4D space, and is valid for all observers. GPS systems prove this point beyond any reasonable doubt.

And like the idea of time existing as a "temporal dimension," which therefore implies the existence of an unobservable "temporal coordinate," the idea of a hypothetical "internal observer" and "external observer" is also unobservable and so also has no direct experimental evidence, and so therefore should be carefully re-examined. Moreva's and others' prediction of a hypothetical observer external to the universe that has access to the abstract coordinate of time is questionable. In their 2014 Physical Review paper, Moreva and his co-authors wrote: "In super-observer mode (Fig. 2a, yellow box) the experimenter takes the place of a hypothetical observer external to the universe that has access to the abstract coordinate time and tests whether the global state of the universe has any dependence on it. Hence, he must perform a quantum interference experiment that tests the coherence between the different histories (wavefunction branches) corresponding to the different measurement outcomes of the internal observers, represented by the which-way information after the polarizing beam splitter PBS1" (Moreva et al., 2014). According to our opinion, the introduction of an observer that is external to the universe is questionable, because we do not have a single piece of data that confirms that such an observer could actually even exist. Models of space-time, i.e., models in which time is understood to be the fourth dimension of space, have no ability to describe entanglement without the introduction of hypothetical elements, e.g., an "observer which is external to the universe". Once it is understood that universal space is timeless, entanglement can be fully understood without the introduction of hypothetical elements, and therefore questionable elements, such as "coordinate time," "inner observer," or an "observer which is external to the universe". Recent research confirms that universal space, which has its origin in a fundamental three-dimensional quantum vacuum, is the direct information medium of EPR-type experiments (Fiscaletti, Sorli, 2017).

On the other hand, in Martinetti (Martinetti, 2013) suggested that the problem of time lies in explaining its emergence, both in quantum mechanics as an idealized abstract parameter in the background space of the observables of the system, as well as in relativity as a geometrical parameter of four-dimensional space-time. Our view of time as the sequential numerical order of material changes in the timeless space of Now allows a resolution of this problem in a clear and unifying way, as follows: both in relativity and in quantum mechanics time is a mathematical parameter measuring the numerical order of material changes in the timeless space of Now. And if in Borghi (Borghi 2016) claims that there is a fundamental inequivalence between thermal clocks and atomic clocks, in our approach it is not allowed – as it seems to be by Borghi – that there is a fundamental physical disagreement between the measurements obtained by relativistic clocks and those provided by thermal clocks in the same experimental situations. In our approach, physical time exhibits a different nature and different features depending on the level of observation of phenomena, which cannot be merged into an ultimate unifying theory, because, at a fundamental level, time existing as a primary physical reality of its own does not exist. Thus, in our opinion, Borghi's argument is poorly expressed. In our approach, the fundamental fact regarding the interpretation of the problem of atomic clocks and thermal clocks lies in nothing more than the existence of two kinds of time; namely, fundamental time, which is the numerical order of changes and exists independent of the observer, and emergent time, which is the duration of material changes and originates from the observer's measurement. One can say that Borghi's argument regarding the possible (eventual) disagreements between thermal clocks and atomic clocks in fact deals with the level of emergent time, but not the level of fundamental time. At the most fundamental level, both for atomic and thermal clocks, time is measured as the sequential numerical order of material changes in the timeless space of Now.

In summary, in light of the arguments made in this paper, one can conclude that the fundamental arena of natural processes is a timeless background, and that material changes run in the timeless space of Now. This means that, when change X_2 comes into existence, change X_1 no longer exists; when change X_3 comes into existence, change X_2 no longer exists. Changes are irreversible, in the sense that a change that no longer exists is lost forever. Any increase in the entropy of a given system only happens in timeless space. In other words, "time flow" and "arrow of time" have only a mathematical existence. Time (as the sequential numerical order of changes) flows within the timeless space of Now, and has an arrow which points in the direction of numerical order increase: $X_1, X_2,...,X_n$.

In Newtonian physics, the fundamental arena of the universe is a space in which time runs uniformly, and so is a fundamental arena in which material changes run in time. Conversely, in Einstein's relativistic physics, time no longer runs in space, because in relativistic physics, time has become the fourth dimension of space in which material changes run. In this article we have shown that time is a mathematical parameter of changes that run in timeless space, bringing together into a unifying picture ideas that have already been proposed by Girelli, Liberati, and Sindoni, as well as by other authors, such as Prati, Caticha, Elze.

During the period of time in which Einstein came up with his Relativity theory, what we now know to be the linear psychological time of "past-present-future" was still considered to be an actual physical reality, and so the ability of this idea to influence physics remained strong. As a result, the idea of linear time as being the fourth dimension of space was incorporated into physics. In this way, time was sort of half-way integrated into space, somewhat like the way a round peg can be sort of half-way integrated into a square hole if one pounds on it hard enough.

In this article however, we are no longer trying to pound the round peg of time into the square hole of space, so to speak. To the contrary, in this article, rather than trying to merge space and time as some sort of composite physical reality, we have shown that time, at a fundamental level, is merely the sequential numerical order of material changes, i.e. motions, that run in space. In this way, time is finally, fully, and accurately integrated into the timeless 4D space of Now: 3D space and time T (Newtonian physics) → 3D+T space-time (Relativity) → 4D timeless space (this article).

In Relativity it is accepted that motion from point A to point B in space happens in 3 spatial dimensions and one temporal dimension. In this article however, we have shown that, in relativity regime, motion happens in 4 spatial dimensions, and we have also shown that that motion happens exclusively in timeless space. In GPS systems clocks run simultaneously in timeless space. Their "relativistic rate" depends on SR and GR effects and is valid for all observers. As time is not an actual physical dimension, no clock can "tick" in some hypothetical physical past or future, since past and future, as has also been described in this article, are purely conceptual realities, and as such have no demonstrable physical correlate. All clocks in this universe run in the same timeless space, which means they all only ever run Now. GPS systems have proven that the relative rate of clocks is valid for all observers. This means that an observer on a train-station platform where there is a stationary clock, and an observer in a passing train where there is moving clock, will each observe the individual clocks to run at the same rate, i.e., they will be able to agree that the clock on the train, which is moving, is running at a somewhat slower rate than the clock on the platform. In this famous example of Special Relativity, it needs to be understood that the stationary observer and the stationary clock, as well as the moving observer and the moving clock, all exist, and so all run, in timeless Minkowski 4D space.

4. **Conclusions**

In 20th-century physics, time is understood as a fundamental physical reality in which the universe exists. However, as of the beginning of the 21st-century, we still have not obtained a single piece of data that provides any evidence whatsoever confirming time to have an actual physical existence. Understanding time to have only a mathematical existence provides a plausible solution to Einstein's Now conundrum: time is the mathematical parameter of changes i.e. motions, which run in the timeless space of Now.

References:

Albert Einstein Site Online http://www.alberteinsteinsite.com (1915)

Anderson, E.: "Machian classical and semiclassical emergent time", arXiv:1305.4685v2 [gr-qc] (2013).

Andrew Jaffe, "Physics: Finding the time", *Nature* **537**, 616 (29 September 2016)

Borghi C., "Physical time and thermal clocks", Foundations of Physics 46, 1374-1379 (2016).

Caticha, A.: "Entropic dynamics, time and quantum theory", *Journal of Physics A: Mathematical and Theoretical* **44**, 22, 225303 (2011); e-print arXiv:1005.2357v3 [quant-ph].

Catalin V. Buhusi, Warren H. Meck, What makes us tick? Functional and neural mechanisms of interval timing, *Nature Reviews Neuroscience* 6, 755–765 (2005)

Gödel, K., "An Example of a New Type of Cosmological Solutions of Einstein's Field Equations of Gravitation," *Rev. Mod. Phys.* **21**, 447, published July 1, 1949

Góźdź, A. and Stefanska, K.: "Projection evolution and delayed choice experiment", *Journal of Physics: Conference Series* **104**, 012007 (2008).

Hawking, S. W. (1992). "Chronology protection conjecture". *Phys. Rev. D.* **46**: 603.

Elze, H. T.: "Quantum mechanics and discrete time from "timeless" classical dynamics", *Lecture Notes in Physics* **633**, 196 arXiv:gr-qc/0307014v1 (2003a).

Elana Gomel, Postmodern Science Fiction and Temporal Imagination, Continuum International Pulsing Group, 2010

Ellis John, N.E. Mavromatos, D.V. Nanopoulos, Derivation of a vacuum refractive index in a stringy space–time foam model, Physics Letters B 665 (2008) 412–417

Ekaterina Moreva, Giorgio Brida, Marco Gramegna, Vittorio Giovannetti, Lorenzo Maccone, and Marco Genovese, Time from quantum entanglement: An experimental illustration, Phys. Rev. A 89, 052122 – Published 20 May 2014

Florian Girelli, Stefano Liberati, and Lorenzo Sindoni, Is the Notion of Time Really Fundamental? Symmetry 2011, 3, 389-401;

Fiscaletti, D. & Sorli, A. Searching for an adequate relation between time and entanglement, Quantum Stud.: Math. Found. (2017) 4: 357.

Fiscaletti, D., Sorli, A.: Perspectives of the numerical order of material changes in timeless approaches in physics. Found. Phys. **45**(2), 105–133 (2015)

Fiscaletti D., Sorli A., and Klinar D., The symmetrized quantum potential and space as a direct information medium", Annales de la Fondation Luois de Broglie 37, 41-71 (2012).

Hermann Minkowski, "Space and Time" in Hendrik A. Lorentz, Albert Einstein, Hermann Minkowski, and

Hermann Weyl, *The Principle of Relativity: A Collection of Original Memoirs on the Special and General Theory of Relativity* (Dover, New York, 1952) pp. 75-91.

Licata, I.: "Minkowski Space-Time and Dirac Vacuum as Ultrareferential Reference Frame", *Hadronic Journal* 14, 3 (1991).

Masanori Sato, Gravitational effect on the refractive index: A hypothesis that the permittivity, ε0, and permeability, μ0 are dragged and modified by the gravity, https://arxiv.org/vc/arxiv/papers/0704/0704.1942v3.pdf (2007).

Martinetti, P. "Emergence of time in quantum gravity: is time necessary flowing?", Kronoscope 13, 67-84 (2013).

Neil Ashby, Relativity and the Global Positioning System, Living Rev Relativ. 2003; 6(1): 1.

Prati, E.: "The nature of time: from a timeless Hamiltonian framework to clock time of metrology", arXiv:0907.1707v1 [physics.class-ph] (2009).

Palle Yourgrau, A World Without Time: The Forgotten Legacy of Gödel and Einstein, Basic books 2006

Page Don N. and Wootters William K., Evolution without evolution: Dynamics described by stationary observables, Phys. Rev. D 27, 2885 – Published 15 June 1983

Richard A. Muller, "Now: The Physics of Time", Amazon 2016.

Rovelli, C.: "Loop Quantum Gravity", *Living Reviews in Relativity*, 1, 1, DOI: 10.12942/lrr-1998-1, http://relativity.livingreviews.org/Articles/lrr-1998-1/ (1998).

Rovelli, C.: "A new look at loop quantum gravity", *Classical and Quantum Gravity* 28, 11, 114005 (2011); e-print arXiv:1004.1780v1 [gr-qc].

Shapiro Irwin I. (1964). "Fourth Test of General Relativity". Physical Review Letters. **13** (26): 789–791

Stockum, W. J. van (1937). "The gravitational field of a distribution of particles rotating around an axis of symmetry.". Proc. Roy. Soc. Edinburgh. 57.

Advanced Relativity: Reintroduction of absolute Frame of Reference

Abstract

In Relativity theory, 'space' is empty and has no physical properties. Motion of a given physical object in such an empty space can be related only to the other objects which exist in space. In Advanced Relativity, space has Planck energy density which is diminished by the presence of a given physical object. But since relativistic motion of physical objects causes 'The Dragging effect' with space, which opens the possibility for the reintroduction of space as the absolute frame of reference.

Key words: relativistic motion, space, absolute reference frame, Relativity Theory, inertia, gravity

1. **Introduction**

Alice and Bob are travelling in the vastness of universal space by the two ships. The spaceships are coming closer and Alice informs Bob that the speedometer in his spaceship does not work. Bob informed Alice that the speedometer is also damaged.

Both Alice and Bob want to know which spaceship has higher speed. Both of them are experts in General Relativity and they decide to calibrate their clocks when driving one by the other and drive a circle with radius of 1000 kilometres. They read their clocks at a reunion. The clock of Alice is ticking slower than Bob's clock. This is according to the relativity, by means of which the Alice's spaceship is moving faster than Bob's spaceship. The rate of clocks in both Alice and Bob's spaceships are defined by the speed of their spaceships, in the same way as the clocks of GPS system have different rate on the satellite and Earth surface. The rate of clock on the satellite is slower than on the Earth surface by 7 microseconds per day (Sorli, 2018).

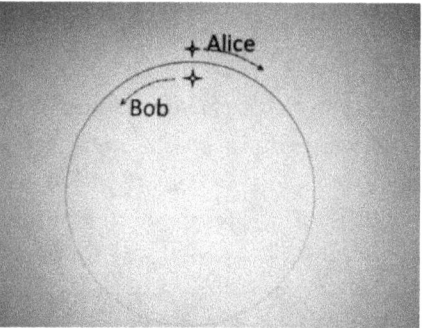

Figure 1: Alice and Bob defining which spaceship is faster

Alice's spaceship diminishes energy density of space more than Bob spaceship because of its speed. Higher is the value of diminished energy density of space, slower is the rate of clocks. This means that energy density of space defines the rate of clocks. Clocks are devices with which we can indirectly measure the values of space energy density in a given point of universal space. In intergalactic space where there are no stellar objects, rate of the clock is at the maximum and energy density of space is at the maximum, which is of Planck energy density ρ_{PE}. On the surface of stellar objects rate of the clock will diminish because energy density of space there is diminishing too. Energy density of space is diminishing also in the centre of spaceship (ρ_{SE}) which is moving with velocity v for the following value:

$$\rho_{SE} = \rho_{PE} - \left(\frac{m \cdot c^2}{V}\right) - \left(\frac{m \cdot v^2}{2V}\right) \quad (1),$$

where m is the mass of the spaceship and V is the volume of the spaceship. Kinetic energy of the spaceship $m \cdot v^2/2$ causes additional diminishing of space energy density in the centre of spaceship and slower rate of clocks in the spaceship regarding rate of clocks on the Earth surface (Sorli, 2018).

2. Discussion

In this perspective, space is the absolute frame of reference for moving of the physical objects. In a given physical object, the rate of clock is at maximum, when object is at rest, regarding the space in which physical object exists. The rate of clock diminishes with the increasing of object velocity. Theoretically, the velocity of a given spaceship can be defined by the rate of clock placed in the spaceship, which is far away from the stellar objects. When spaceship is at rest regarding the space, rate of clock is at the maximum and is diminishing by the increased velocity of the spaceship.

Historically it's seen that abolishing of the absolute frame of reference has happened with the abolishing of ether in the beginning of 20th century. Einstein has managed to describe the constancy of light speed in all inertial systems by introducing space-time of Minkowski with Lorentz transformation. But in

Advanced Relativity, the photon is the vortex of space (Sorli, 2018) and has the same velocity in all the inertial systems. In this view, the Special Relativity can be described in a 3D Euclidean space with Galilean transformation for spatial coordinates X, Y and Z and Selleri transformation for time t (Fiscaletti, Sorli, 2013; Sorli et al. 2011). In this description there is no need for introducing the "proper time", "coordinate time", and "length contraction. Relative velocity of clocks in all inertial systems in Advanced Relativity valid for all observers, which is proved by GPS system and can be expressed by Selleri's formula, in which there is no spatial coordinates. Below is the Lorentz formula for time dilation, where X is spatial coordinate of inertial system with time t:

$$t' = \frac{t - \frac{v \cdot X}{c^2}}{\sqrt{1 - \frac{v^2}{c^2}}} \quad (2),$$

Lorentz formula for time dilation was further developed by Selleri as follows below:

$$t' = t \cdot \sqrt{1 - \frac{v^2}{c^2}} \quad (3),$$

where we can see that time t' and so rate of clock in inertial system X',Y',Z' depends only on the velocity v of moving inertial system X',Y',Z'.

In Advanced Relativity, the space is not filled with ether. It's well known that the space itself has physical property, namely 'variable energy density'. Photon is the vortex of space and moves with the velocity of light speed c. In all different inertial systems with all different velocities, the light satisfies the law of Doppler effect. The first postulate of Special Relativity (constancy of light speed) has physical origin in photon as the vortex of space in which all inertial systems move. The second postulate of Special Relativity (physical laws are the same in all inertial systems) has physical origin in the fact that all inertial systems move in the same space which energy density is defines physical laws. The Relativity of time has origin in variability of space energy density and is valid for all observers.

Both the rest and moving photon clocks have the same rate, because light has the same velocity in all inertial systems. It is not true that the rate of the photon clock is slower for the stationary observer. Because of this optical illusion, the moving photon clock will not have slower rate; stationary observer has no 'magic powers' to change the rate of the moving clock.

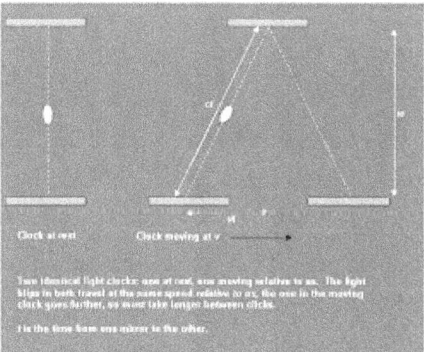

Figure 2: Moving photon clock has slower rate because of optical illusion of stationary observer

2.1. Relative and Absolute velocity

You imagine you drive a car with the velocity of 110 km per hour. 500 meters in front of you is a car moving with the velocity of 120 km per hour. Accident happens and the car crash in the wall along the street. Driver luckily survives but the car is totally destroyed. This simple example proves that your "relative velocity" of 10 km per hour of the car in front of you was not "real". The real velocity of the car is the velocity regarding the highway which is 120 km per hour.

The same is valid per light speed which "real" velocity is the velocity of the light in the 4D space and is valid for all observers. If you move towards the source of light with the velocity of 1000 km per hour it make no sense to say that for you light has velocity of v = c + 1000 km/h.

In Advanced Relativity, both the inertia and gravity are the result of dynamics between given massive object and the space. The massive object is diminishing energy density of space accordingly to the amount of its energy, which creates both the inertial and gravitational mass, as below

$$E = mc^2 = (\rho_{PE} - \rho_{SE}) \cdot V \qquad (4),$$

where E is energy of space, m is mass of the object, V is volume of the object, and $(\rho_{PE} - \rho_{SE})$ is the difference between Planck energy density of outer space ρ_{PE} and energy density of space in the centre of physical object ρ_{PE}. Difference $(\rho_{PE} - \rho_{SE})$ is the origin of inertia and gravity (Sorli, 2018). So that we have

$$E = (\rho_{PE} - \rho_{SE}) \cdot V \qquad (5),$$

which clearly shows that the energy of a given massive physical object has origin and corresponds to both the 'difference of energy density of space outside' and the 'centre of the massive physical object'. Equations (4) and (5) are expressing fundamental symmetry in the universe between a given material object and space in which this object exits. We have to understand that the space does not end on the surface of material object; rather it continues inside. Elementary particles are the structures of space energy, space and particles are in the final stage "the same stuff" as was already said by Ervin Schrödinger: *"What we observe as material bodies and forces are nothing but shapes and variations in the structure of space"* (Sorli et al., 2018a).

In General Relativity the gravitational time dilation is calculated by the following formula:

$$t = \frac{t_0}{\sqrt{1 - \frac{2GM}{rc^2}}} \qquad (6),$$

where t_0 is rate of the clock on the surface of stellar object surface, M is the mass of stellar object, G is gravitational constant, r is radius of the stellar object, and t is the rate of the clock at the point T which is infinitely away in empty cosmic space. For example, when one second has passed at point T in infinity, only 0.9999999993 seconds has passed on the Earth clock. So that we can calculate the rate of clock at the point T_1, situated on the distance h above the surface of stellar object with following formula:

$$t = t_0 \cdot \sqrt{\frac{1 - \frac{2GM}{(r+h) \cdot c^2}}{1 - \frac{2GM}{rc^2}}} \qquad (7).$$

Let's calculate time t at the point 20 km above Earth surface comparing with the 1 day elapsed time on the Earth surface:

$$t = 86400s \cdot \sqrt{\dfrac{1 - \dfrac{2(5.97219 \times 10^{24} kg)(6.67408 \times 10^{-11} m^3 kg^{-1} s^{-2})}{(6371000m + 20000m)(8.99 \times 10^{16} m^2 s^{-2})}}{1 - \dfrac{2(5.97219 \times 10^{24} kg)(6.67408 \times 10^{-11} m^3 kg^{-1} s^{-2})}{(6371000m)(8.99 \times 10^{16} m^2 s^{-2})}}}$$

$$t = 86400s \cdot \sqrt{\dfrac{1 - \dfrac{7.9717748 \times 10^{14} m^3 s^{-2}}{(6391000m)(8.99 \times 10^{16} m^2 s^{-2})}}{1 - \dfrac{7.9717748 \times 10^{14} m^3 s^{-2}}{(6371000m)(8.99 \times 10^{16} m^2 s^{-2})}}}$$

$$t = 86400s \cdot \sqrt{\dfrac{1 - 0.00000000138747}{1 - 0.00000000139183}}$$

$$t = 8.6400000000188352 \times 10^4 s \quad (20 \text{ km above surface})$$

Let's calculate time t at the point 40 km above Earth surface comparing with the 1 day elapsed time on the Earth surface:

$$t = 86400s \cdot \sqrt{\dfrac{1 - \dfrac{7.9717748 \times 10^{14} m^3 s^{-2}}{(6411000m)(8.99 \times 10^{16} m^2 s^{-2})}}{1 - \dfrac{7.9717748 \times 10^{14} m^3 s^{-2}}{(6371000m)(8.99 \times 10^{16} m^2 s^{-2})}}}$$

$$t = 86400s \cdot \sqrt{\frac{1-0.00000000138315}{1-0.00000000139183}}$$

$$\Delta_{space.energy.density} = \rho_{PE} - \rho_{SE} = 6.138 \cdot 10^{19} \, J/m^3$$

$$t = 8.6400000000374967 \times 10^4 \, s \quad \text{(40 km above surface)}$$

Let's calculate time t at the point 60 km above Earth surface comparing with the 1 day elapsed time on the Earth surface:

$$t = 86400s \cdot \sqrt{\frac{1 - \dfrac{7.9717748 \times 10^{14} \, m^3 s^{-2}}{(6431000m)(8.99 \times 10^{16} \, m^2 s^{-2})}}{1 - \dfrac{7.9717748 \times 10^{14} \, m^3 s^{-2}}{(6371000m)(8.99 \times 10^{16} \, m^2 s^{-2})}}}$$

$$t = 86400s \cdot \sqrt{\frac{1-0.00000000137884}{1-0.00000000139183}}$$

$$t = 8.6400000000561158 \times 10^4 \, s$$
(60 km above surface)

Let's calculate time t at the point 80 km above Earth surface comparing with the 1 day elapsed time on the Earth surface:

$$t = 86400s \cdot \sqrt{\frac{1 - \dfrac{7.9717748 \times 10^{14} \, m^3 s^{-2}}{(6451000m)(8.99 \times 10^{16} \, m^2 s^{-2})}}{1 - \dfrac{7.9717748 \times 10^{14} \, m^3 s^{-2}}{(6371000m)(8.99 \times 10^{16} \, m^2 s^{-2})}}}$$

$$t = 86400s \cdot \sqrt{\frac{1-0.00000000137457}{1-0.00000000139183}}$$

$$t = 8.6400000000745632 \times 10^4 s \quad \text{(80 km above surface)}$$

In Advanced Relativity, we can calculate the energy density of the space at the point T by the following formula:

$$\rho_{SE} = \rho_{PE} - \frac{3M \cdot c^2}{4\pi \cdot (r+d)^3} \quad (8),$$

where ρ_{SE} is energy density at the point T, ρ_{PE} is Planck energy density, r is diameter of the stellar object, and d is the distance between the centre of the stellar object and point T. Calculation below shows that the diminishing of the energy density of space on the Earth surface regarding Planck energy density ρ_{PE}. It's noted that when $d = r$, one gets energy density of space ρ_{SE} in the centre of the stellar object as follows in formula below:

$$\rho_{SE} = 4.633 \times 10^{113} J/m^3 - \frac{3(5.97219 \times 10^{24} kg)(299792458 ms^{-1})^2}{4\pi \cdot (1.2742 \times 10^7 m)^3}$$

$$\rho_{SE} = 4.633 \times 10^{113} J/m^3 - \frac{(1.791657 \cdot 10^{25} kg)(8.99 \times 10^{16} m^2 s^{-2})}{4\pi (2.66877 \times 10^{21} m^3)}$$

$$\rho_{SE} = 4.633 \times 10^{113} J/m^3 - \frac{1.6107 \times 10^{42} m^2 s^{-2}}{2.6 \times 10^{22} m^3}$$

$$\rho_{SE} = 4.633 \times 10^{113} J/m^3 - 6.1949 \times 10^{19} J/m^3$$

$$\Delta_{space.energy.density} = \rho_{PE} - \rho_{SE} = 6.1949 \times 10^{19} J/m^3.$$

Again, calculation below also shows that the diminishing of the energy density of space 20 km above the Earth surface regarding Planck energy density:

$$\rho_{SE} = 4.633 \times 10^{113} J/m^3 - \frac{3(5.97219 \times 10^{24} kg)(299792458 ms^{-1})^2}{4\pi \cdot (6371000m + 6391000m)^3}$$

$$\rho_{SE} = 4.633 \times 10^{113} J/m^3 - \frac{(1.791657 \times 10^{25} kg)(8.99 \times 10^{16} m^2 s^{-2})}{4\pi (2,0785296 \times 10^{21} m^3)}$$

$$\rho_{SE} = 4.633 \times 10^{113} J/m^3 - \frac{1.6107 \times 10^{42} m^2 s^{-2}}{2.6119574 \times 10^{22} m^3}$$

$$\rho_{SE} = 4.633 \times 10^{113} J/m^3 - 6.166 \times 10^{19} J/m^3$$

$$\Delta_{space.energy.density} = \rho_{PE} - \rho_{SE} = 6.166 \times 10^{19} J/m^3.$$

The calculation below shows that the diminishing of the energy density of space 40 km above the Earth surface regarding Planck energy density:

$$\rho_{SE} = 4.633 \times 10^{113} J/m^3 - \frac{3(5.97219 \times 10^{24} kg)(299792458 ms^{-1})^2}{4\pi \cdot (6371000m + 6411000m)^3}$$

$$\rho_{SE} = 4.633 \times 10^{113} J/m^3 - \frac{1.6107 \times 10^{42} m^2 s^{-2}}{2.6242566 \times 10^{22} m^3}$$

$$\rho_{SE} = 4.633 \times 10^{113} J/m^3 - 6.138 \times 10^{19} J/m^3$$

$$\Delta_{space.energy.density} = \rho_{PE} - \rho_{SE} = 6.138 \times 10^{19} J/m^3.$$

Similarly, calculation below shows that the diminishing of the energy density of space 60 km above the Earth surface regarding Planck energy density:

$$\rho_{SE} = 4.633 \times 10^{113} J/m^3 - \frac{3(5.97219 \times 10^{24} kg)(299792458 ms^{-1})^2}{4\pi \cdot (6371000m + 6431000m)^3}$$

$$\Delta_{space.energy.density} = \rho_{PE} - \rho_{SE} = 6.109 \times 10^{19} J/m^3.$$

Calculation below shows the diminishing of the energy density of space 80 km above the Earth surface regarding Planck energy density:

$$\rho_{SE} = 4.633 \times 10^{113} J/m^3 - \frac{3(5.97219 \times 10^{24} kg)(299792458 ms^{-1})^2}{4\pi \cdot (6371000m + 6451000m)^3}$$

$$\rho_{SE} = 4.633 \times 10^{113} J/m^3 - \frac{1.6107 \times 10^{42} m^2 s^{-2}}{2.6242566 \times 10^{22} m^3}$$

$$\rho_{SE} = 4.633 \times 10^{113} J/m^3 - 6.080 \times 10^{19} J/m^3$$

$$\Delta_{space.energy.density} = \rho_{PE} - \rho_{SE} = 6.080 \times 10^{19} J/m^3.$$

	4,633x10E113	
Earth surface	-6,192x10E19	86400 s
20 km above	-6,166x10E19 26	+0,000000 188352 s
40 km above	-6,138x10E19 54	+0,000000 374967 s
60 km above	-6,109x10E19 83	+0,000000 561158 s
80 km above	-6,080x10E19 112	+0,000000 745632 s

Figure 3: Calculations of gravitational time dilation and of increasing of energy density of space

We can see in figure 3 above that by increasing the distance from the Earth surface time dilation is increasing, and the energy density of space is also increasing. On the basis of these calculations we can draw the following result (figure 4). For time dilation, we apply the last numbers 188, 374, 561, 745. For energy density increasing we apply decimal differences between space energy at Earth surface and 20km which is 26, 40km which is 54, 60km which is 83, 80 km which is 112.

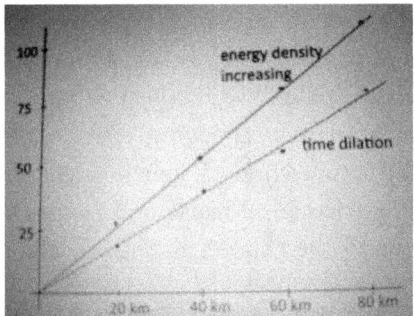

Figure 4: Diagrams of gravitational time dilation and space energy density

We can see that diagram of time dilatation and diagram of increasing energy density of space have similar linear shape and direction which confirms that the actual physical circumstance which determines the rate of the clock is energy density of space in which is always NOW. It's well known that motion happens and clocks run in space only; not in time. Time is merely a numerical sequential order of events running in timeless space. When we measure numerical order time as duration appears. Duration of a given observed phenomenon enters the existence when measured from the side of the observer (Sorli et al., 2018b).

In Advanced Relativity, the space and time are finally separated again after more than 100 years. Space is fundamental energy of the universe and time is only a mathematical parameter of motion in space. Motion does not happen in time and space, motion happens only in space and time is just its numerical sequence. Space and a given physical object are not separable elements of the universe. A given physical object is diminishing energy density of space. Rotation of physical object or stellar object causes also rotating of surrounding space, which we call "dragging effect". Sun is rotating space around it and this rotation of space causes precession of the planets. We develop a model of precession of planets which is based on dragging effect: axial rotation of physical objects causes minimal rotation of 4D space. Dragging effect was derived in 1918, in the framework of general relativity, by the Austrian physicists Josef Lense and Hans Thirring, and is also known as the Lense-Thirring effect.

Figure 5: Dragging effect

Our calculations on planets precession got exact the same result as Einstein's calculations. Because of its rotation 4D superfluid space is pushing a bit the planets and so their perihelion is every year a bit ahead as it should be according to Newton physics.

Our model also explains 'Sagnac effect' which can't be explained by the classical Relativity with empty space. When we measure the signal, which moves from A to B on the Earth surface in the direction of Earth motion, the elapsed time will be shorter than when we measure the time of signal moving from B to A. This happens because the space around the Earth is rotating with the Earth. As signal is a vibration of space, it needs lesser time, while moving from A to B.

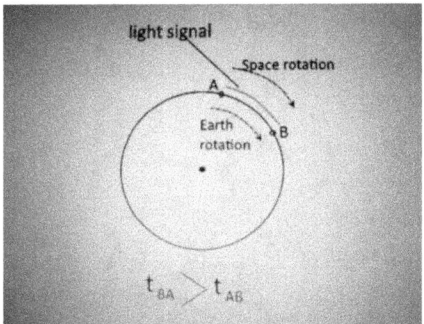

Figure 5: Sagnac effect in Advanced Relativity

Michelson-Morley experiment has given null result because it is carried out on the preposition that Earth is moving in stationary ether. In Advanced Relativity, this mistake in improved by understanding that surrounding space is moving with the Earth and is rotating with the Earth. That's why Michelson-Morley experiment has given null result (Fiscaletti, Sorli, 2016). It is not that the energy of ether is filling the space, as it was thought in 19th century. Space itself is a fundamental energy of the universe. Einstein's Relativity is marvellous "mathematical theory" based on the idea of space-time and is developed in Advanced Relativity in "physical theory" where space gains concrete physical properties, namely, energy density which gives origin to inertia and gravity.

In the leading paradigm of today mainstream physics we have in space three different fundamental fields: electromagnetic fields, Higgs field and gravitational field. These three fields are not unified yet. In Advanced Relativity the functions of these three fundamental fields are carried by the space itself: photon is the vortex of space composed out of magnetic and electric component, inertia and gravity have both origin in the difference $\rho_{PE} - \rho_{SE}$ between Planck energy density of outer space ρ_{PE} and energy density of space in the centre of physical object ρ_{PE}. Introduction of energy density of space in physics is necessary for the development of physics.

3. Conclusions

The idea of empty space is one the biggest misunderstandings of 20th century physics. Space has energy density which is interacting with physical entities from the micro to the macro scale. For the motion of every physical object, space is the absolute reference frame. When a clock in space is at rest and far away from stellar objects, it has a maximum speed. When the clock in space as the absolute frame of reference starts mowing, its rate is diminishing. GPS system is working on the basis of different rates of clocks on the Earth surface and on the satellites, which are determined by the variable energy density of space. Kinetic energy of the moving physical object is the energy of space, which is additionally concentrated in the object. Space is fundamental non-created energy of the universe.

References:

Fiscaletti D., Sorli A., About a new suggested interpretation of special theory of relativity within a three-dimensional Euclid space, Annales Physica, Vol. 68 (2013)

Fiscaletti D., Sorli A., Dynamic Quantum Vacuum and Relativity, Annales Physica, Vol. 71 (2016)

Sorli A., Fiscaletti D., Klinar D., New insights into the special theory of relativity, Physics Essays, Vol. 24, Num. 2 (2011)

Sorli A., Bijective Analysis of Physical Equations and Physical Models, NeuroQuantology, Vol. 16, Num. 6 (2018)

Sorli A., Dobnikar U., Patro S.K., Mageshwaran M., Fiscaletti D., Euclidean-Planck Metrics of Space, Particle Physics and Cosmology, NeuroQuantology, Vol. 16, Num. 4 (2018a)

Sorli A., Kaufman S., Fiscaletti D., Minkowski Space-time and Einstein's NOW Conundrum, NeuroQuantology, Vol. 16, Num. 5 (2018b)

Epistemology Crisis of Today's Physics

Abstract

In today physics it often happens that experimental data is interpreted as the proof of the phenomenon that has not been directly observed and is based only on the theoretical model. With the obtained data, the model becomes recognized as "real" and the phenomenon that the model describes also become recognized as "real" – it is acknowledged as the physical reality, although it has not been observed by instruments or human senses. This situation is leading physics into deep epistemology crisis, which is not yet seen today. The aim of this article is stressing the situation and finding the solution to overcome the crisis.

Key words: Epistemology, Higgs field, gravitational waves.

1. **Introduction**

Recently, two Nobel prizes have been given for the discovery of phenomena which have not been observed by instruments or human senses, namely: Higgs field and gravitational waves. In the Higgs field research, it was measured that extremely rarely (one in milliard collisions of protons) is measured a characteristic flux of energy named "Higgs boson". The discovery of the Higgs boson in today physics means the proof for the existence of Higgs field which was not measured or observed by human senses. In gravitational waves research, it was measured that the laser light motion in the LIGO interferometer needs sometimes (when gravitational wave is supposed to pass the interferometer) a bit longer or shorter time when passing the beams. The minimal time variability of the laser light is, in today physics, the proof for the existence of gravitational waves which were not measured or observed by human senses. This "epistemological gap" between obtained data and their interpretation represents a serious problem from the view of the epistemology of physics.

2. A weak point of today's methodology of physics

Special Theory of Relativity (STR), published in 1905, has deeply changed the methodology of physics. STR has caused that still today it is fully acknowledged that time is the 4th dimension of space which has physical existence. The formalism $X4 = ict$ has convinced the majority of physicists that time is the 4th dimension of space-time model. Later on, the acceptation of the space-time model has convinced physicists that time as the 4th dimension of space has real physical existence, although there is no experimental evidence for this and, moreover, nobody has ever seen time as the 4th dimension of space. With STR mathematical model it has happened that mathematics has overruled physics. If today you have a consistent mathematical model, every peer review journal will publish it, without bothering about the epistemological stability of the article. Epistemological stability means the level of the adequacy of the model with physical reality. The most epistemologically stable would be the model which is related to the physical reality with the bijective function of set theory. The bijective function prescribes to each element a physical reality as exactly one element in the model:

$$f : X \to Y \quad (1).$$

Recent research where the so-called "bijective epistemology" was applied has confirmed that the model of space-time has no "bijective epistemological stability". Time is not the 4th physical dimension of space (Fiscaletti and Sorli, 2015). It seems that Einstein was aware of the fact that time has an exclusively mathematical existence and he added the imaginary number i in the formula $X4 = ict$. After a certain time i was removed from the formula, which then became $X4 = ct$. At the end constant c was written as 1 and formula has taken the form $X4 = t$, which was then the exact mathematical description of the conviction that material changes are running in some real physical time as the 4th dimension of space. This conviction has no epistemological stability, it is the biggest theoretical failure of the 20th century physics and, however, it is still valid in the mainstream physics today.

The weak point of today's physics methodology is that, based on mathematical models which are proved by *indirect experiments* (indirect experiments are the ones which do not directly measure the phenomenon that is the subject of a given research), the conclusions are taken about the existence of these phenomena as if they had an actual physical existence. The best examples are the existence of Higgs field and gravitational waves, which both have not been directly observed and measured. They have been only theoretically predicted and described with the mathematical model, for which we do not know how much it correspondents to the physical reality.

3. A historical overlook of how mathematics has overruled physics

Length contraction was postulated by George FitzGerald (1889) and Hendrik Antoon Lorentz (1892) to explain the negative outcome of the Michelson–Morley experiment and to rescue the hypothesis of the stationary aether (Lorentz–FitzGerald contraction hypothesis) (*FitzGerald, George Francis, 1889*). Length contraction was later adopted by Einstein and used in his STR and also in his General Relativity Theory (GRT).

The idea that the length of a given moving object could change its length has no minimal common sense. When the idea was mathematically described, it has become plausible because, at that time, nobody did care much about the negative epistemological consequences of such a methodology.

In today physics, mathematical models are playing the decisive role. If you are a theoretical physicist and you have a mathematically consistent model of a given phenomenon for which you predict it could exist, nobody will ask you about the epistemological stability of your model. On the contrary, an experimentalist will try to prove your model with some indirect experiment.

Let us go back to the end of the 18th century and imagine that we are actively participating in the ether research. We cannot see the ether directly with our senses. We can only see the light which we suppose is the wave of the ether. Our idea is that ether is not stationary, ether is moving with the objects. We can imagine the planet Earth moving through the ether in the way that ether which is around the Earth is moving and rotating with the Earth. Ether we cannot strictly divide from the physical object, they are intrinsically bonded and should be examined together. Such imagination is rescuing the ether hypothesis without the introduction of the length contraction. Photon is the wave of ether and behaves according to the Doppler effect. Every inertial system is moving in the ether and light which is the wave of ether has the same velocity in every inertial system. When the *distance between the source of the light and the inertial system that is shortening the light will increase frequency, when the distance is increasing, the light frequency will decrease. Ether which is surrounding the Earth is rotating with the Earth. We call this phenomenon ether drift (in today physics it is called "quantum vacuum dragging effect") and it fully explains and mathematically describes Sagnac effect, which STR cannot explain (Fiscaletti and Sorli, 2016).*

In physics, it happens that wrong imaginations are summed up. The idea of stationary ether has caused that ether was thrown out of the physics. Having no more the medium of light, Einstein has created the idea that a photon can move in an empty space deprived of physical properties. This idea is the cause of the crisis in today's physics, in which Standard model tries to describe physical reality only with the fields and particles which exist in an empty space. We have, in today physics, three main fields (electromagnetic quantum vacuum of quantum electrodynamics QED, Higgs field and gravity field) and we are not able to recombine these fields in a unique model.

Advanced Relativity model is staying with the ether under the new name "dynamic quantum vacuum", where photon is the wave of quantum vacuum. Inspired by the work of Max Planck, the ether is given a physical property of Planck energy density, which has minimal variations according to the mass of a given physical object: diminished energy density of ether (quantum vacuum) corresponds to the mass of a given physical object:

$$E = mc^2 = (\rho_{PE} - \rho_{qvE}) \cdot V \quad (2),$$

where ρ_{PE} is Planck energy density, ρ_{qvE} is energy density of 4D superfluid space in the centre of a given particle (or massive body), m is mass of the particle (or massive body), V is volume of the particle (or massive body) (Sorli et al.,2017). Advanced Relativity model works perfectly without Higgs field and without gravitational field. Mass and gravity both originate in the variable energy density of quantum vacuum.

4. Introduction of bijective epistemology in physics

Bijective epistemology requires that the phenomenon which is examined needs to be observed with the senses of the observer or detected with an instrument. Human perception and instrumental perception are the obligatory elements for the beginning of the research on a given subject. When the research subject is predicted theoretically, that is, based on an existent model, it needs to be confirmed with a direct human observation or direct instrumental measurement. A good example of this is Dmitri Mendeleev's research, who published the first periodic table of the chemical elements in 1869, based on properties that appeared with some regularity, as he laid out the elements from the lightest to the heaviest (Kaji, 2002). When Mendeleev proposed his periodic table, he noted gaps in the table and predicted that as-then-unknown elements existed with properties appropriate to fill those gaps. Unknown elements have been later discovered by different researchers.

The discoveries of Higgs field and gravitational waves are epistemologically weak, because they have not been measured directly. Today, they are recognized as big achievements of physics. If this will still be the case in 2117, it is doubtful. There is no doubt that the periodic system of elements will be still valid in 2117, because it is based on direct measurements. Indirect measurements applied in the research of Higgs field and gravitational waves should be carefully examined before their full application can be used as the standard in future research. The question seems philosophical, but it is not. It touched the core of physics and deserves attention of both theoretical physicists and experimentalists.

5. **The realization of Einstein's vision of completeness of a theory**

Bijective epistemology is fulfilling Einstein's vision of "completeness" of a theory": "If, without in any way disturbing a system, we can predict with certainty (i.e., with probability equal to unity) the value of a physical quantity, then there exists an element of physical reality corresponding to this physical quantity." And for a theory to be complete, "every element of the physical reality must have a counterpart in the physical theory" (Bernstein 1999). In Advanced Relativity, every element in the model corresponds to exactly one element in the physical reality.

Einstein used to say: "Imagination is more important than knowledge. For knowledge is limited, whereas imagination embraces the entire world, stimulating progress, giving birth to evolution." We must add here that scientific imagination, in order to lead us to coherent models, needs to be based on human perception and experimental data. If not, imagination can be developed in a mathematical model (with no real correspondence with the physical world), for which we then tirelessly search until we confirm it with an indirect measurement. We are proposing a new research methodology in physics, which is fully allowing creative imagination and is based on human perception and experimental data.

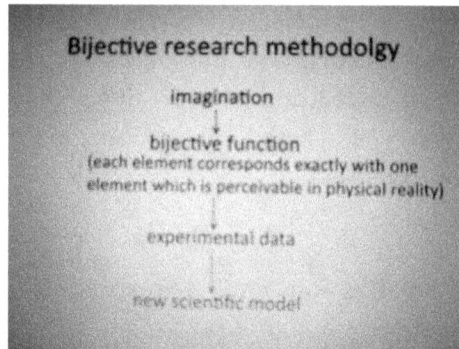

Figure 1: Bijective research methodology

Bijective research methodology is excluding the possibility of an error in the process of scientific research in physics. This methodology is giving more credibility to the creative imagination based on perception, rather than to pure mathematic speculation, which is often disconnected from the physical world. Higgs mechanism, for example, is based on pure mathematics speculation and is as such epistemologically unstable.

6. Discussion

Advanced Relativity is based on bijective research methodology. Advanced Relativity has kept the ether as the physical basis of the universal space. Ether is not in the space; ether is the "stuff" out of which space is made. Ether is dynamic in the sense that ether is moving with the objects and is rotating with them. We call ether with the new name "quantum vacuum". In the end of the 18th century, light was understood as the wave of the ether. We cannot see and detect ether directly; we can see and detect light as its vibration. Advanced Relativity has adopted this view because Einstein's idea that photon could move in an empty space deprived of physical properties is more philosophical than scientific: light needs a physical medium in which light moves. This physical medium must also be an element of the scientific model of the physical world. Between 4D superfluid space as the element of physical reality and the 4D space in the model of Advanced Relativity there is a bijective function:

$$f : X \to Y \quad (1),$$

where X represents 4D physical superfliud space and Y represents in the model of 4D space of Advanced Relativity.

Advanced Relativity is built on bijective methodology and has following advances:

1. In the universal space, there is always NOW. Linear time belongs to the mind. With clocks, we measure the duration of material changes, i.e. motions in space.
2. No signal can move in time, every signal can move in space only and time is the duration of its motion. CMBR radiation is the radiation of quantum vacuum, where there is always NOW. Big Band Theory has come to the end.
3. In the space velocity of light, it has minimal changes and depends on the energy density of quantum vacuum. "Gravitational time dilation" actually means that light minimally diminishes the speed, because in stronger gravity, energy density of quantum vacuum is a bit lower, which changes its permittivity and permeability, and so velocity of light. We know from

classical physics that density of the medium defines velocity of the signal. The denser the medium, the faster the motion of the signal. This is also valid in Advanced Relativity.
4. Mass and gravity both have origin in variable energy density of quantum vacuum.
5. Dark energy is the energy of quantum vacuum.
6. In Advanced Relativity, there is no "inner observer", "outer observer", "coordinate time", "proper time". Velocity of clocks in all inertial systems is valid for all observers and does not depend on them.
7. Relative velocity of clocks in inertial systems depends only on the variable energy density of quantum vacuum and is valid for all observers. GPS proves that clocks on the satellites and on the Earth surface run with the same velocity for all observers. It if would not have been be so, GPS could not work (*Fiscaletti and Sorli, 2016*).
8. *Curvature of space in GRT is the mathematical description of variable energy density of quantum vacuum. Space is not "curved" in a physical sense (Fiscaletti and Sorli, 2014). NASA confirms universal space is "flat"; it has Euclidean form (NASA, 2013).*

7. Conclusions

Epistemology crisis of today physics has roots in the conviction that the development of new mathematical models will develop physics. This is true only partially because mathematical models of today physics are often the result of pure speculation, which is not grounded in human perception and experimental data. To overcome the crisis, this article has proposed a new methodology, which allows only imagination based on human perception and experimental data.

References:
Bernstein, H.J. Foundations of Physics (1999) 29: 521. https://doi.org/10.1023/A:1018856024112

Fiscaletti D. and Sorli A., SPACE-TIME CURVATURE OF GENERAL RELATIVITY AND ENERGY DENSITY OF THERE DIMENSIONAL QUANTUM VACUUM, Annales UMCS Sectio AAA: Physica, VOL.LXV. (2014).

Fiscaletti D and Sorli A., Bijective epistemology and space-time. Foundations of Science; 20(4): 387-398. (2015)

Fiscaletti D. and Sorli A., Dynamic Quantum Vacuum and Relativity. Annales UMCS Sectio AAA: Physica, VOL. LXXI. (2016).

FitzGerald, George Francis (1889), "The Ether and the Earth's Atmosphere", *Science, **13** (328): 390.*

Kaji, Masanori (2002). "D. I. Mendeleev's concept of chemical elements and *The Principles of Chemistry*". *Bulletin for the History of Chemistry.* **27** (1): 4–16.

NASA, http://map.gsfc.nasa.gov/universe/uni_shape.html (2013).

Sorli A., Kaufman S., Dobnikar U., Fiscaletti D., Advanced Relativity for the Renaissance of Cosmology and Evolution of Life, NeuroQuntology, Vol. 15, Num. 4 (2017)

Advanced Relativity for the Renaissance of Cosmology and Evolution of Life

Abstract

In Advanced Relativity model conscious observer builds the cosmological model on the basis of elementary perception and experimental data. Each element of the model is related to the exactly defined one element of the universe with the bijective function of set theory. The physical universe represents a set X; the cosmological model represents a set Y. Both sets are related with the bijective function of set theory $f : X \to Y$. The resulting theoretical model of the universe is an adequate picture of the universe as a non-created system in dynamic equilibrium in which life is the consistent part of cosmic dynamics.

Key words: conscious observer, bijective function, epistemology, cosmology, evolution of life

1. Introduction

Conscious observer is building cosmology model exclusively on his elementary perception and experimental data. Imagine you are a conscious observer. You have the perfect ability to observe and be conscious how your mind is building scientific models of reality. With your eyes you can observe in the universe matter, energy and space. Matter and energy are continuously changing, so you observe material changes of the universe. These material changes are characterized by an exact sequential order: change X_2 is happening after change X_1 and change X_3 is happening after change X_2. When change X_2 enters existence change X_2 does not exist anymore. When change X_3 enters existence, change X_2 does not exist anymore. Material changes in the universe have their numerical order which you observer observe with the sight and measure with the clocks. This numerical order of material changes is time. These are the 5 fundamental elements of the universe which you perceive with eyes and experience without interference (interpretation, analysis, synthesis) of your mind: matter, energy, space, change and time. And you, as the conscious observer, are the sixth fundamental element of the universe. These six elements constitute the set X of the universe:

$$X : \{M_X, E_X, S_X, C_X, T_X, CO_X\} \quad (1),$$

where M_X is matter, E_X is energy, S_X is space, C_X is change, T_X is time, CO_X is observer.

You may apply bijective function of the set theory:

$$f : X \rightarrow Y \quad (2),$$

and define the set Y of the model of the universe with the following elements:

$$Y : \{M_Y, E_Y, S_Y, C_Y, T_Y, CO_Y\} \quad (3),$$

where M_Y is matter, E_Y is energy, S_Y is space, C_Y is change, T_Y is time, CO_Y is observer.

Set X and set Y are related with bijective function which realizes the Einsteinian idea of "completeness theorem" according to which each element of the model should correspond exactly to one element of the real world (Fiscaletti and Sorli, 2015a).

Conscious observer has experimental data that black holes have tendency to become smaller and smaller and that finally disappear. He calculates that in black holes energy density of space becomes lower and this makes atoms unstable. On the basis of these calculations he predicts that in black holes matter transforms into the photons which then turn back into the energy of space (Sorli et al., 2016a). This means that matter, energy (electromagnetic energy) and space are different forms of energy of the set X and forms the subset energy X (EX), which can be formulated as following:

$$X : \{CO_X, C_X, T_X, \{XE\}\} \quad (4)$$

$$EX : \{M_X, E_X, S_X\}.$$

We have now four fundamental elements of the set universe X: conscious observer, change, time and energy; the same is in the set model Y:

$$Y : \{CO_Y, C_Y, T_Y, \{YE\}\} \quad (5)$$

$$YE : \{M_Y, E_Y, S_Y\}.$$

Conscious observer has experimental data that space is continuously radiating cosmic rays which form elementary particles (Friedlander, 2002). In the universe matter is continuously transforming back into energy of space in black holes; in outer space energy of space is continuously transforming into cosmic rays and further into elementary particles. This means that energy of the universe is continuously circulating. The question of the "begging" of the universe does not exist for the conscious observer. He understands that universe is not running in some linear time which exists only as a mind frame in which common observer is experiencing run of universal changes. Universe runs in space where is always NOW. Conscious observer is exceeding the 20th century view on the universe as a system which has its beginning in some temporal remote event called Big Bang.

In physics of 20th century the principle of causality was one of the most valid. No one ever had doubt that this principle could be wrong in a sense that it does not reflect the real nature of physical reality. The core of the causality principle is that events are running in time and that event 1 is a cause of event 2, event 2 is a cause of event 3 and so on. Advanced Relativity has confirmed that events run only in space and that time is their numerical order. When event 2 enters into existence, event 1 does not exist anymore, when event 3 enters into existence, event 2 does not exist anymore. Events run in space where there is always NOW. The principle of causality which implies the existence of physical time as the arena of the events does not work in physical reality; it is only our rational mind interpretation.

In Advanced Relativity principle of causality is replaced with the principle of dynamics: Universe is in a continuous dynamics which does not have particular cause. Dynamics itself is the fundamental universal principle. Dynamics always happens between two or more elements. For example motion of an elementary particle in space is dynamics between particle and space. In Advanced Relativity there are two different types of dynamics in the universe:

1. Temporal dynamics
2. Immediate dynamics.

Temporal dynamics is for example, motion of planets and stellar objects. Motion has its own numerical order and is temporal. It does not run in time, it runs in space only, time is their numerical order which we measure with clocks. Immediate dynamics has no numerical order. Such dynamics are gravity and entanglement.

Universal phenomena based on dynamics do not need energy. Motion of the Moon around Sun does not require any energy use, because it is the result of dynamics between variable energy density of space and two material objects. In physics of 20th century the prevalent idea was that for anything to happen energy is used. This is valid only for manmade machines and is not valid for the universe which is ruled by the law of dynamics. Dynamics is the intrinsic physical property of the universe and nature which does not use energy in order to run. For common view of physics this seems "strange", however our perception of the universe is confirming the principle of dynamics. Planets are rotating around the Sun without using energy; the same is with the rotation of our solar system around the centre of Milky way galaxy. Cosmologists today still approach the universe similarly to the machines made by the humans. This "temporal cosmological view" is based on causality principle where cause and consequence are happening in some linear physical time. In Advanced Relativity, time is merely the numerical parameter of motion in space where is always NOW which means that cause and consequence happen in the same NOW and are part of universal dynamics which can be immediate or temporal.

In Advanced Relativity universal space has origin in quantum vacuum. We developed formula for gravitational constant G in order to show that G is related to permittivity and permeability of quantum vacuum.

$$G = \frac{l_P^3}{m_P \cdot t_P^2} \quad (6).$$

$$G = \frac{l_P^3}{m_P \cdot \frac{l_P^2}{c^2}} \quad (7).$$

$$G = \frac{c^2 \cdot l_P}{m_P} \quad (8).$$

Velocity of light c is:

$$c = \frac{1}{\sqrt{\omega_0 \varepsilon_0}} \quad (9).$$

$$c^2 = \frac{1}{\omega_0 \cdot \varepsilon_0} \quad (10).$$

Combining equations (8) and (10) we get:

$$G = \frac{l_P}{m_P \cdot \omega_0 \cdot \varepsilon_0} \quad (11).$$

Equation (11) shows gravitational constant G depends on Planck metrics of quantum vacuum, its permittivity and its permeability.

Max Planck units are representing mathematical values for quantum vacuum physical properties. In empty space energy density of quantum vacuum has a value of Planck energy density ρ_{PE}. We know in physics that every physical system has a tendency to reach the average distribution of energy. Where a given particle exists, quantum vacuum energy density is smaller exactly for the amount of the energy contained in a given particle:

$$\rho_{PE} = \rho_{qvE} + \frac{mc^2}{V} \quad (12),$$

where ρ_{PE} is Planck energy density, ρ_{qvE} is energy density of quantum vacuum in the centre of a given particle (or massive body), m is mass of the particle (or massive body), V is volume of the particle (or massive body).

Equation (12) we can rearrange and we will get:

$$E = mc^2 = (\rho_{PE} - \rho_{qvE}) \cdot V \quad (13)$$

where left side of the equation represent famous Einstein equation and right side represents the missing part which explain origin of energy and mass of a given particle or massive body (Sorli, 2017). Equation (13) is the "supersymmetry formula" which shows that the energy E and the mass m of a given particle are made out of the same "stuff" called quantum vacuum and that energy and mass are symmetric to the diminishing of the energy density of quantum vacuum ρ_{qvE} in the centre of a given particle. Formula (13) is also valid for massive objects and stellar objects; it works from micro to the macro level of the universe.

Formula (13) can also be multiplied by Lorentz factor and is then valid for energy E_R of relativistic particles, which because of their speed absorb energy of the quantum vacuum, which becomes their kinetic energy:

$$E_R = \gamma E = \gamma mc^2 = \gamma(\rho_{PE} - \rho_{qvE}) \cdot V \quad (14).$$

Equations (13) and (14) are presenting the origin of mass of elementary particles without the introduction of Higgs mechanism. According to the mass-energy equivalence, no field can exist which would give mass (mass here means energy) to the particles. The only plausible idea would be that the interaction of particles with some field generates their inertial mass. "Mass" means the amount of energy incorporated in a given particle and "inertial mass" means the particle property to stay at the certain position or to move in the certain direction. In Advanced Relativity, the inertial mass of elementary particles and massive bodies have the origin in diminished energy density of quantum vacuum. The dynamics between diminished energy density of quantum vacuum and a given massive object (particle, massive object or stellar object) generates inertial mass and gravitational mass (Sorli, 2017).

High-speed motion of particles and massive bodies creates friction with the quantum vacuum. This friction causes the concentration of quantum vacuum energy in the relativistic object which actually is its kinetic energy. In CERN Collider are colliding about hundred millions of pairs of relativistic protons each second. Only one in ten billion of collisions create so called "Higgs boson" which is nothing more than the released relativistic energy of two relativistic protons. The conclusion that the Higgs boson is the proof of the existence of Higgs field (Higgs boson is the "ripple" of the Higgs field) which should give particles mass is epistemologically unstable. Might happen, the next generation of physicists will re-evaluate the discovery of "God particle", and give it appropriate meaning and importance.

2. Weak points of Big Bang cosmology

According to the Big Bang cosmology, universe is happening in space-time as the fundamental arena of the universe. Space-time has 3 spatial dimensions and 1 temporal dimension. According to Bijective epistemology presented in the section 1 above, the element of space-time cannot have existence in the set model universe Y where time is merely numerical order of changes in space. Element of space S_Y belongs to the energy subset YE and element of time T_Y is not an element of the energy subset YE. In this view space and time cannot be united in space-time. Space-time is a model which has no solid epistemological ground and thinking that it has some correspondence in real world, seems incorrect.

Big Bang should happen in some remote physical past. The prove for that should be cosmic microwave background radiation CMBR which has origin in Big Bang and is radiating since the Big Bang via time as the 4th physical dimension of space. Bijective epistemology does not allow a given signal to move in time because time has only a mathematical existence. Every signal moves only in space and time (fundamental time) is the numerical order of its motion. When fundamental time is measured by the observer emergent time which is the duration enters into existence (Fiscaletti and Sorli, 2015b). In Advanced Relativity CMBR has origin in a fundamental universal background space which can be also called 3D quantum vacuum and is defined by reduction-state (**RS**) processes of creation/annihilation of particles/antiparticles (with opposite orientations of spins), corresponding to elementary fluctuations of the quantum vacuum energy density.

Big Bang cosmology model uses finite spherical Riemann geometry. NASA results have confirmed universal space is flat and corresponds to the Euclidean geometry (NASA, 2016). This means that use of Riemann geometry in cosmology is not allowed. Universe is infinite in its dimensions and has also infinite amount of the energy:

$$E_U = \infty \quad (15).$$

Calculations which are taking in account energy and mass of the universe should be finite seems not appropriate. Universe cannot be approached as a finite system, universe is infinite. We can study the dynamics of the universe in the region of universal space which is reachable with the telescopes and hope that universe which is beyond our observation functions according to the same laws as the observable universe.

Big Bang cosmology predicts universe is finite and is expanding. We have seen above universal space is infinite and infinite space cannot expand; we could only speculate that the observable area of the universe is expanding. The main proof for expanding universe should be red shift where light of the galaxies is moving to the red spectrum because of their motion away from the Earth since universe is expanding. Expansion of observable universe is questionable because "red shift" can also be interpreted as a consequence of light pulling from the strong gravity (Pound, 2000). This so called "gravitational red shift" is a basis for "tired light" hypothesis of Swiss astronomer Fritz Zwicky. He proposed that the reddening effect was not due to motions of the galaxy, but to an unknown phenomenon that caused photons to lose energy as they travelled through space. He considered the most likely candidate process to be a drag effect in which photons transfer momentum to surrounding masses through gravitational interactions; and proposed that an attempt be made to put this effect on a sound theoretical footing with general relativity. He also considered and rejected explanations involving interactions with free electrons, or the expansion of space (Zwicky, 1929).

3. Evolution of life is a component of universal cosmic dynamics

Life, as we know it, beyond a common range of compatibility linked to the thermo dynamical conditions and to the chemistry of a set of elements (carbon, hydrogen, azotes, oxygen, phosphorus, sulphur), exhibits an extraordinary variety which cannot be reduced to a simple chemistry-physics. Despite life being a physical phenomenon, subjected to the general laws which rule the behaviours of matter and energy, this compatibility is not enough to reproduce the variety of the living systems: biological systems satisfy a sort of indifference principle in the sense that are historic systems and can exhibit unforeseeable behaviours, which cannot be classified (Licata, 2010). According to the Maturana and Varela autopoiesys theory in living systems it is the global functional dynamics to fix each time the boundary conditions and the back-reaction cycles which support their autonomy (Maturana and Varela, 2001). Living processes seem to be characterized by the so-called intrinsic emergency, associated with the appearance of properties which are compatible with the models describing the basic relations between system and environment, but absolutely unforeseeable because, in similar situations, several variations are possible. The appearance of these new emergent properties can modify in an irreversible way the nature of the system and its relations with the environment (Licata, 2015).

All the evidence indicate that organism and environment are intimately interconnected, from the socio-cultural domain right down to the genes. Stable inheritance depends on this very interconnection, rather than on a mythical unchangeable genome. The process of heredity has a dynamic stability that resides in the feedback interrelationships that can propagate from the external environment through the physiological system to the genes. Organism and environment engage in ceaseless rounds of mutual definition and transformation, which is the essence of evolution. Cycles of feedback between the biosphere and the physicochemical environment are the basis of stability for the global ecosystem.

Organisms may be interconnected with one another and with their physicochemical environment by information flow, as well as by material and energy flow. Most molecular biologists assume that the answer to biological organization will come when all the molecules in organisms are isolated and analysed. But biological organization is a dynamic, macroscopic order extending over astronomical numbers of molecules, spanning distances at least millions of times the size of individual molecules (Mae-Wan Ho, 2016). This organization enables organisms to transform energy with the rapidity and efficiency rarely achieved elsewhere and to be extremely sensitive to specific signals in the environment.

In 1960, Nobel laureate biochemist Albert Szent-Györgyi pointed out that we can begin to understand the characteristics of living systems only if we take into account the collective properties of molecules akin to superconductivity and superfluidity. This idea was developed at about the same time by German-born British solid-state physicist Herbert Fröhlich, who suggested that living systems have collective modes of activity somewhat similar to superconductors operating at physiological temperatures (Szent-Györgyi, 1960). The Fröhlich model implies that, when the energy supply goes above a certain level, the polar structure enters into a state of nonlinear vibration and a coherent behaviour of excited electrons observed in living systems emerges which is similar to coherent behaviours found in superconductors. The only difference is that in superconductors, this behaviour is observed with the help of Bose-Einstein condensation at temperatures near the absolute zero point (Fröhlich, 1968), while coherence in biological systems occurs at room temperature (Reimers *et al.*, 2009). The Fröhlich model suggests that metabolic energy, instead of being lost as heat, is stored in the form of collective or coherent electromechanical and electromagnetic excitations. These "coherent excitations" could be responsible for generating and maintaining long-range order. They also make possible highly efficient energy transfer and transformation of energy and the detection of very weak electromagnetic signals.

Evidence for the existence of coherent excitations in living systems comes from the work of German biophysicist Fritz-Albert Popp and his co-workers (1988), who showed that practically all organisms and cells emit light (biophotons) at very weak intensities. Organisms and cells also re-emit light at higher intensities as delayed luminescence after exposure to a brief pulse of light. As the result of some 15 years of experimental work, Popp became convinced that biophotons come from a coherent electro-dynamical field within the living system. This field has a wide range of frequencies that are coupled together to give effectively a single degree of freedom, and that may be the basis of biological organization. Living systems are thus both emitters and receivers of electromagnetic signals originating from the physicochemical environment as well as from other organisms.

In order to explore the biological organization which characterizes living systems, in the recent paper "The unified spacememory network: from cosmogenesis to consciousness", authors consider the possibility that nonlocal information dynamics, intrinsic to the properties and behavior of material systems and uniquely harnessed by the natural nanotechnology of supramolecular systems of the brain (similar to the Hameroff-Penrose model of orchestrated objective reduction) are involved in producing the sentience, awareness, and memory of cognitive processes (Haramein *et al.*, 2016). Moreover, they propose that nonlocal influences across spatial and temporal domains, communicated through the micro-wormhole network of the Planck-scale geometric structure of spacetime, may play an instrumental role in the evolution and development of physical systems, thus engendering an ordering dynamic as well as directionality towards higher levels of complexity and organizational synergy. In this approach, one has the same unifying ordering dynamics regarding the evolution and development both of physical systems and of biological systems and therefore awareness properties follow the same rules describing the evolution of matter and of the universe. One has a unifying picture of evolution and development processes, from non-organic matter to biological matter, which are shaped by the integrative and ordering influences of a holographic Planckian wormhole network, from cosmogenesis to the universe being aware of itself, i.e. consciousness. In Haramein's, Brown's and Val Baker's model living systems are seen as complex systems composed by biomolecules intercommunicating in an intricate network of information and energy exchange. These highly complex biomolecules are constituted in turn by elements that evolve under the influence of a larger, more fundamental intercommunication system, where, at each scale, there are dynamics creating order and directionality of interaction towards higher levels of synergetic organization, and as a result, greater levels of consciousness.

On the other hand, cosmic energy may be connected and thus forms a matrix within the entire cosmos and by means of structured matter can elicit life (Penrose *et al.*, 2011). Evolution began much after the existence of energy and matter and its unanimity and therefore the answer to the origin of life lies much before the emergence of viruses, bacteria, algae and eukaryotes, which required the presence of water and organic compounds. For life to emerge there had to be a right blend of energy and matter, where the properties of energy and matter played an important role. Many significant studies seem to suggest that, as regards the problem of the origin of life, the properties of matter and energy are not emergent, but must be considered as the designers for creation (Maysinger *et al.*, 2015). Moreover, recently, theories associated with redox homeostasis have been considered as an important aspect linked to the origin of life (Allen, 2010). On the other hand, quantum computation has unknowingly opened up a new array of hope towards understanding the origin of life from the perspective of organic and inorganic chemistries and the use of non-living matter to perform activities like living matter. Several theories are now available that support the emergence of consciousness from quantum based mechanisms (Hameroff and Penrose, 2014) and the involvement of the cosmic energy uptake in the form of electromagnetic radiation (Pereira, 2015a; McFadden, 2007 and 2013). A recent pathway known as the cell-soul pathway has been proposed to be a hypothetical mechanism where a single cell uses the quantum phenomenon to convert external cosmic energy to internal energy, to store and use this energy as part of its conservation process (Pereira, 2015b) indicating that life and consciousness is quantum processed and may have driven the origination of living forms proceeding with evolution. In the recent paper "Origin of life: a consequence of cosmic energy, redox homeostasis and quantum phenomenon", Reddy and Pereira suggest that origin of life emerges as an eternal process associated with the interaction between energy from the cosmos and inorganic matter, which supports matter with retention of this riveted energy, as energy to be circulated within the primitive channelized structures to conserve energy by the materialization of the proton homeostasis mechanisms developed from the obtainable inorganic matter (Reddy and Pereira, 2016). According to Reddy's and Pereira's approach, origin of life is therefore a result of the organization and reorganization of matter to support constants such as the cosmic energy, matter and quantum processes that prevailed in the cosmos and mellowed with evolution.

In the light of the results of Maturana, Varela, Szent-Györgyi, Fröhlich, and the most recent ones of Popp, Haramein, Brown and Val Baker, Reddy and Pereira, the authors of this paper suggest the possibility that evolution of life is a consistent part of cosmic dynamics, is a continuation of the evolution of the universe. In Advanced Relativity (AR) model evolution of life is the integral part of universal dynamics. Life is developing in entire universe and is tending to create intelligent and conscious organisms as we humans are. In Advanced Relativity (AR) the fundamental primordial space of the universe is consciousness which is described by the n-dimensional Hilbert space. Entire universe exists in consciousness and the manifestation of consciousness is of varying degrees and such manifestations happen due to the intervention of pure cosmic energy/Prana (Jayakrishnan *et al.*, 2017). That is why matter has tendency to develop in life and life further in conscious organisms. Life is the subsystem of the universe and cannot be examined separately from the system in which it develops, which is the universe. This view of Advanced Relativity is supported by a classical Vedic text, viz., Sri Saundarya Lahari written by Adi Sankaracharya, the first verse reveals the nature and existence of the universe, i.e., the universe is a manifestation of pure consciousness and pure cosmic energy cannot be separated from pure consciousness (Jayakrishnan et al., 2017).

Advanced Relativity is unifying cosmology and biology. Evolution of life seen from the view of conscious observer is the process which runs in the universe. Universe is a system in a permanent dynamic equilibrium and as totality has no entropy. Increasing of entropy of the matter that we observe in the universe is only a part of universal dynamics. All over the universe matter has tendency to decrease entropy because matter is existing in space which is syntropic energy (Sorli et al., 2016b). This decreasing entropy tendency can be observed as the presence of all organic molecules essential for development of life in entire universal space. This so-called "chemical evolution" then develops in living organisms on the planets with similar physical circumstances as on the planet Earth. Consciousness is shaping life on 3D physical reality via pilot photons of higher dimensional Hilbert spaces (Sorli et al., 2017).

In Advanced Relativity the notion of energy can be extended into multidimensional Hilbert spaces. Energy can be defined in a nD Hilbert space where n is the cardinal number of natural numbers; nD Hilbert space is the fundamental background of consciousness, which acts in the human being as the observer. This means that life on the molecular level of 3D dimensionality has origin in nD consciousness. Evolution of life is encoded in 4D and more dimensional Hilbert spaces of higher information density and is communicated to the 3D ordinary life dimensionality via bio-photons.

4. Conclusions

Planet Earth is the subsystem of the solar system which is the subsystem of Milky way which is the subsystem of the universe. Approaching evolution of life as a process which has developed on the planet Earth surface is a narrow viewpoint having its root in the geocentric model according to which Earth is the center of the universe. The conscious observer is dropping geocentric model and building his scientific picture of the universe and life on the elementary perception and bijective epistemology. The result of the conscious observer research methodology is the model of the stationary universe which is a non-created system in a permanent dynamic equilibrium. Evolution of life on planet Earth is an integral part of cosmic dynamics which is developing in the entire universe. Evolution of the universe and evolution of life should not be approached separately because they are one phenomenon. Attained knowledge of today's science has reached to the point where cosmology and evolution of life can be unified in the new discipline called "Cosmo-biology".

References:

Allen JF. Redox Homeostasis in the Emergence of Life. On the Constant Internal Environment of Nascent Living Cells. J Cosmol 2010; 10: 3362-3373.

Fiscaletti D and Sorli A. Bijective epistemology and space-time. Foundations of Science 2015a; 20(4): 387-398.

Fiscaletti D and Sorli A. Perspectives of the Numerical Order of Material Changes in Timeless Approaches in Physics. Foundations of Physics 2015b; 45(2): 105-133.

Friedlander MW. A Thin Cosmic Rain: Particles from Outer Space, Harvard University Press, Harvard, 2002.

Fröhlich H. Long-Range Coherence and Energy Storage in Biological Systems. Int. J. Quant. Chem. 1968; 2:641–649.

Hameroff S and Penrose R. Consciousness in the universe. A review of the 'Orch OR' theory. Phys Life Rev 2014; 11: 39–78.

Haramein N, Brown W.D. and Val Baker A. The unified spacememory network: from cosmogenesis to consciousness. Neuroquantology 2016; 14(4):657-671.

Jayakrishnan TP, Sorli A and Nair RR. Numerical and consciousness: How hot is the bond?. NeuroQuantology 2017 (In press).

Licata I. There's plenty of hidden information at the quantum bottom, in *New trends of quantum information*, edited by Licata I, Sakaji A, Felloni S and Singh JP, Aracne Editore, Rome, 2010.

Licata I. I gatti di Wiener. Riflessioni sistemiche sulla complessità, Bonanno Editore, Acireale-Rome, 2015.

Mae-Wan Ho, Meaning of life and the universe, World Scientific, Singapore, 2016.

Maturana H and Varela F. Autopoiesi e cognizione. La realizzazione del vivente, Marsilio Edizioni, Venice, 2001.

Maysinger D, Ji J, Hutter E and Cooper E. Nanoparticle-based and bioengineered probes and sensors to detect physiological and pathological biomarkers in neural cells. Front Neurosci 2015; 9 (480). doi: 10.3389/fnins.2015.00480.

McFadden J. Conscious electromagnetic field theory. NeuroQuantology 2007; 5(3):262-270.

McFadden J. The CEMI Field Theory Closing the Loop. J Conscious Stud 2013; 20:153–168.

NASA, http://map.gsfc.nasa.gov/universe/uni_shape.html 2013. Accessed Date, June 2016.

Penrose R, Hameroff S, Stapp H and Chopra D. Consciousness and the Universe: Quantum Physics, Evolution, Brain and Mind, Cosmology Science 2011.

Pereira C. The cosmic energy bridge, cellular quantum consciousness and its Connections. J Metaphysic Connect Conscious 2015a.

Pereira C. Electromagnetic Radiation, a Living Cell and the Soul: A Collated Hypothesis. NeuroQuantology 2015b; 13(4): 426-438. doi: 10.14704/nq.2015.13.4.862.

Popp FA., Li KH, Mei WP, Galle M and Neurohr R. Physical Aspects of Biophotons. Experientia (1988); 44:576–585.

Pound RV. Weighing Photons. Classical and Quantum Gravity 2000; 17: 2303–2311. doi: 10.1088/0264-9381/17/12/301.

Reddy JK and Pereira C. Origin of life: a consequence of cosmic energy, redox homeostasis and quantum phenomenon. Neuroquantology (2016); 14(3):581-588.

Reimers JR, McKemmish LK, McKenzie RH, Mark AE, Hush NS. Weak, Strong, and Coherent Regimes of Fröhlich Condensation and their Applications to Terahertz Medicine and Quantum Consciousness. Proc. Natl. Acad. Sci. U. S. A., 2009; 106: 4219–4224.

Sorli AS. Mass-Energy Equivalence Versus Higgs Mechanism. American Journal of Modern Physics. Special Issue: Insufficiency of Higgs Mechanism 2017; 7(4-1):1-4. doi: 10.11648/j.ajmp.s.2018070401.11

Sorli A, Dobnikar U, Fiscaletti D and Koroli V. Advanced Relativity: Multidimensionality of Consciousness and Mind, Origin of Life, PSI Phenomena, NeuroQuantology 2017; 15(2): 109-117.

Sorli A, Fiscaletti D and Mageshwaran M, Advanced Relativity: Unification of Space, Matter and Consciousness. NeuroQuantology 2016b; 14(4): 645-656.

Sorli A, Koroli V, Nistreanu A and Fiscaletti D. Cosmology of Einstein's NOW. American Journal of Modern Physics. Special Issue: Insufficiency of Big Bang Cosmology 2016a; 5(4-1): 1-5. doi: 10.11648/j.ajmp.s.2016050401.11

Szent-Györgyi A. Introduction to a Sub-Molecular Biology. Academic Press, New York, 1960.

Zwicky F. On the Red Shift of Spectral Lines through Interstellar Space, Proceedings of the National Academy of Sciences 1929; 15(10): 773-779, Bibcode: 1929PNAS...15..773Z, doi: 10.1073/pnas.15.10.773, PMC 522555, PMID 16577237.

Advanced Relativity: Unification of Space, Matter and Consciousness

Abstract

Advanced Relativity (**AR**) is the model which is fully integrating matter and consciousness applying Hilbert spaces. Advancer Relativity is reviewing the key concepts of Einstein's Relativity, namely, space and time. In AR space is what we measure with roads and time is what we measure with clocks. Space is not empty and deprived of physical properties; it is characterized by a variable energy density which gives origin to energy, mass and gravity from the micro to the macro scale. Time is merely numerical order of material changes, i.e. motion running in space. AR describes all phenomena of special relativity (SR) and general relativity (GR) and opens new perspectives in cosmology and astronomy, namely, no signal can move in time as time is merely numerical order of a given signal moving in space.

Key words: space, time, consciousness, Hilbert spaces, Special Relativity, General Relativity, super-fluid quantum vacuum.

1. Introduction

Advanced Relativity (**AR**) is the development of Einstein's Relativity Theory which fully integrates the observer [1] and gives curvature of space physical meaning, namely, variable energy density of space. Curvature of space is the mathematical expression of its energy density. Flat space has higher energy density than curved space [2, 3]. Variable energy density of space is origin of energy, mass and gravity of elementary particles [4].

In **AR** universal space has origin in quantum vacuum and time is merely numerical order of changes, i.e. motion. Linear time "past-present-future" belong the mind The Observer in physics not only causes the collapse of wave function when observing a given superposition, but also, when measuring a given physical phenomenon, defines duration as "emergent time". Until measured, time exists only as a fundamental time, which is the numerical order of material changes running in space [5]. In **AR** space is always NOW [6]. In **AR** time is not 4th physical dimension of space, time is merely mathematical parameter of changes which run in space. In **AR** it is not time that is relative, which is the common interpretation of Einstein's Relativity, which is wrong. Time as numerical order of changes in space cannot be relative; relative is velocity of changes which depends on energy density of space. Less dense is space, lower is velocity of changes [7].

The Special Relativity can be described in a 3D Euclidean space. According to the 3D interpretation of special relativity – which can explain in a consistent way several relativistic phenomena, such as the aberration of light, the Doppler effect, Jupiter's satellites occultation and radar ranging of the planets – Einstein's formalism of special relativity based on the standard Lorentz transformations may be derived from a more fundamental 3D Euclidean space, with Galilean transformations for the three spatial dimensions and Selleri's transformation for the rate of clocks [8].

In **AR** light is not bending because space is curved. Light is bending because photon has a mass which is attracted by the gravity of stellar objects. Curvature of space in **GR** is the mathematical description of its variable energy density. Universal space itself is "flat" – Euclidean confirms NASA research [9].

2. Advanced Relativity is without length contraction and without time dilatation

In **AR** we do not have "length contraction" and we do not have "time dilatation". Length contraction was introduced by George FitzGerald and Hendrik Lorentz to explain the negative outcome of the Michelson-Morley experiment and to rescue the hypothesis of the stationary ether. The fact is that Michelson-Morley experiment has given null result because ether around the Earth is moving with the Earth and is also rotating with the Earth. This phenomenon can be called "dragging affect" [3].

When at the beginning of 20th century ether was abolished from physics Einstein has then proposed the idea that photon can move in an empty space deprived of physical properties. This discovery has caused a big theoretical problem, namely, at that time it was observed that light has the same speed independently from the fact that an observer moves from the source or moves towards the source. Considering that photon and the observer move in an empty space Galilean transformation was not able to describe constancy of light. If the model of photon as the wave of ether were not abolished, Galilean transformation would be suitable, namely, the source of light and all inertial systems are moving in ether where photon is the wave of ether. In our model of space as the 4 dimensional superfluid medium photon has the same velocity in all inertial systems because photon is the vortex of the 4D superfluid space which is the medium in which all inertial systems move. When you move towards the source of light or you move away from the souce light is obeying Doppler effect.

Einstein invoked a much more complicated way, namely, constancy of light as a postulate which was not necessary because constancy of light was already measured. In order to describe electromagnetism in an empty space he used Lorentz transformation and Minkowski manifold. The idea of the "length" contraction" was preserved in this model. The problem is that length contraction in SR is valid only for the observer "at rest" and is not valid for the moving observer. How this is possible no one can explain in today physics. Length contraction of SR is illogical. We do not have in physics a single plausible physical explanation why the length is different for rest observer than for the moving observer. It is time to abandon length contraction from physics and to accept that photon is the vortex of 4D superfluid space and is moving in the 4D supertfluid space with the light speed c.

Considering that photon is the vortex of the 4D space in which all inertial systems move **SR** can be described in a 3D Euclidean space with Galilean transformation for X, Y and Z and Selleri's transformation for time t, without "length contraction", "time dilatation" and "time coordinate" [8]. In **AR** it is not time that is relative, relative is velocity of material changes which we measure with clocks. Clocks run in quantum vacuum only, not in time. Time is merely numerical order of changes which run in quantum vacuum which is always NOW [6]. **GPS** proves that relative velocity of clocks because of **SR** and **GR** effect are valid for all observers. "Rest observer" and "moving observer" should be abolished because they do not have real existence; they are only theoretical unproved propositions which are irrational.

Shapiro experiment is interpreted in today physics as a "gravitational time dilatation" which is wrong. The measured fact is that light has minimally smaller speed in strong gravity. Stronger is gravity, lower is density of quantum vacuum, which is changing its permittivity and permeability which causes minimal change of light velocity [3]. In today physics it is not allowed even to think that light speed might have in some special cases minimal variations, which is not reasonable. Infinitesimal small variations of light speed are not calling into question validity of Einstein's Relativity which remains the ground stone of physics and needs only to be renewed in some minimal details.

One of these details is that space-time is a wrong concept, because space-time where time is the 4th dimension of space does not exist in physical reality. Space we measure with roads and time we measure with clocks. Time cannot dilate and space cannot shrink. Time dilatation and length contraction are only mathematical tricks of Einstein in order to solve the problem of constancy of light speed in an empty space. **AR** is solving the constancy of light speed in a much more elegant and non-ambiguous way which is valid for all observers.

AR model of dynamic quantum vacuum describes mathematically in details planets precession where quantum vacuum has a dragging effect because of Sun motion. Our results on precession are on the second the same as of Einstein. Also Sagnac effect is described in details by the dragging effect of rotating interferometer [3].

As Einstein kept length contraction in **SR** he also introduced it in **GR** which is another mathematical trick of Einstein never observed and measured. Gravitational length contraction is based on length variation of beams by LIGO interferometer which also was not measured directly. The only plausible explanation of LIGO results, namely, variable time of laser motion inside the beams is that gravitational waves are changing energy density of quantum vacuum and so its permittivity and permeability which changes minimally light speed [6]. In **AR** gravitational waves are areas of variable energy density of quantum vacuum.

3. Advanced Relativity unifies General Relativity (GR) and Quantum Mechanics (QM)

In **AR** elementary particles and stellar objects both move in space only and not in time. Time is merely numerical order of their motion. Every particle, massive body or stellar object diminishes energy density of space in correspondence to the amount of its energy. Diminished energy density of space can be considered the origin of mass and gravity from the micro to the macro scale.

By starting from a Planck's metric emerging, for example, from loop quantum gravity [10-12], a variable energy density of space can be considered as the fundamental arena giving rise to the different entities and objects existing in the physical universe. In **AR**, in the absence of elementary particles, atoms or massive objects, energy density of space is defined by the Planck energy density given by the following relation:

$$\rho_{pE} = \frac{m_p \cdot c^2}{l_p^3} \quad (1),$$

where m_p is Planck mass, and l_p is Planck length. The energy density of space (1) is the maximum energy density that can be sustained in the minimum quantized space and can be considered as the ground state of the same physical flat-space background. The appearances of material objects and subatomic particles correspond to changes of the energy density of space and thus can be considered as the excited states of the same flat-space background, characterized by a lower energy density than the Planck energy density (1): each excited state of the quantum vacuum is defined by a diminished energy density which corresponds exactly to the energy of the particle under consideration [3].

Each material object endowed with a mass m is produced by a change of the energy density of space on the basis of equation

$$m = \frac{V \cdot \Delta\rho_{qvE}}{c^2} \quad (2),$$

where

$$\Delta\rho_{qvE} = \rho_{pE} - \rho_{qvE} \quad (3),$$

$$\rho_{qvE} = \rho_{pE} - \frac{mc^2}{V} \quad (4),$$

where m is the mass of the object, V is its volume. It must be emphasized here that equations (2)-(4) describe baryonic matter both in macrophysics and in microphysics.

In analogous way to Chiatti's and Licata's transactional approach [13, 14], where the creation and annihilation of an elementary quantum are the two only primary extreme physical events corresponding to a peculiar reduction of a state vector (which is constituted of interaction vertices in which real elementary particles are created or destroyed), in **AR** the appearance of baryonic matter derives from an opportune excited state of the 3D quantum vacuum defined by an opportune change of the quantum vacuum energy density and corresponding to specific reduction-state (**RS**) processes of creation/annihilation of quanta [3]. The excited state of the quantum vacuum corresponding to the appearance of a material particle of mass m is defined (in the centre of that particle) by the energy density (4) (and by the change of the energy density (3), with respect to the ground state) and its evolution is determined by opportune **RS** processes of creation/annihilation of quanta described by a wave function at two components satisfying a time-symmetric extension of the Klein-Gordon quantum relativistic equation

$$\begin{pmatrix} H & 0 \\ 0 & -H \end{pmatrix} C = 0 \quad (5)$$

where $H = -\left(-\hbar^2 \partial^\mu \partial_\mu + \frac{V^2}{c^2}(\Delta \rho_{qvE})^2\right)$.   Equation (5) corresponds to the following equations

$$\left(-\hbar^2 \partial^\mu \partial_\mu + \frac{V^2}{c^2}(\Delta \rho_{qvE})^2\right)\psi_{Q,i}(x) = 0 \quad (6)$$

for creation events and

$$\left(\hbar^2 \partial^\mu \partial_\mu - \frac{V^2}{c^2}(\Delta \rho_{qvE})^2\right)\phi_{Q,i}(x) = 0 \quad (7)$$

for destruction events. By writing the two components of the wave funciton in polar form

$$\psi_{Q,i} = |\psi_{Q,i}| \exp\left(\frac{iS^\psi_{Q,i}}{\hbar}\right) \quad (8),$$

$$\phi_{Q,i} = |\phi_{Q,i}| \exp\left(\frac{iS^\phi_{Q,i}}{\hbar}\right) \quad (9)$$

and decomposing the real and imaginary parts of the Klein-Gordon equation (5), for the real part one obtains a couple of quantum Hamilton-Jacobi equations that, by imposing the requirement that they are Poincarè invariant and have the correct non-relativistic limit, assume the following form

$$\partial_\mu \begin{pmatrix} S^\psi_{Q,i} \\ S^\phi_{Q,i} \end{pmatrix} \partial^\mu \begin{pmatrix} S^\psi_{Q,i} \\ S^\phi_{Q,i} \end{pmatrix} = \frac{V^2}{c^2}(\Delta\rho_{qvE})^2 \exp\begin{pmatrix} Q^\psi_{Q,i} \\ -Q^\phi_{Q,i} \end{pmatrix} \quad (10),$$

while the imaginary part gives the continuity equation

$$\partial_\mu \left(\rho \partial^\mu \begin{pmatrix} S^\psi_{Q,i} \\ S^\phi_{Q,i} \end{pmatrix} \right) = 0 \quad (11)$$

where ρ is the ensemble of particles associated with the spinor under consideration and

$$Q_{Q,i} = \frac{\hbar^2 c^2}{V^2 (\Delta\rho_{qvE})^2} \begin{pmatrix} \frac{\left(\nabla^2 - \frac{1}{c^2}\frac{\partial^2}{\partial t^2}\right)|\psi_{Q,i}|}{|\psi_{Q,i}|} \\ -\frac{\left(\nabla^2 - \frac{1}{c^2}\frac{\partial^2}{\partial t^2}\right)|\phi_{Q,i}|}{|\phi_{Q,i}|} \end{pmatrix} \quad (12)$$

is the quantum potential of the vacuum. The quantum potential of the vacuum is the fundamental entity which the guides the occurring of the processes of creation or annihilation in space and makes the 3D quantum vacuum a fundamentally non-local manifold. In virtue of the primary physical reality of the processes of creation and annihilation and of the non-local action of the quantum potential which is associated with the amplitudes of them (as well as of the opposed sign of its second component with respect to the first component, which seems to indicate that it is not possible to go backwards in the physical time [15]), in the 3D quantum vacuum the duration of the processes from the creation of a particle or object till its annihilation has not a primary physical reality but exists only in the sense of numerical order. In other words, the 3D quantum vacuum, as a fundamental medium subtending the observable forms of matter, energy and space-time, is a timeless background. The behaviour of the matter in the universe can be seen as an undivided network of **RS** processes that take place in the 3D timeless quantum vacuum and time exists merely as a mathematical parameter measuring the dynamics of a particle or object.

As regards the **RS** processes of creation, the quantum potential associated with the virtual particles of the **RS** processes of the 3D quantum vacuum may be written also as

$$Q = V\frac{p_1 + p_2}{n} = -\frac{\hbar^2 c^2 n^2}{4\Delta\rho_{qvE}^2 V^2}\left[\nabla^2 \Delta\rho_{qvE} - \frac{1}{c^2}\frac{\partial^2}{\partial t^2}\Delta\rho_{qvE}\right] +$$

$$\frac{\hbar^2 c^2 n^2}{8\Delta\rho_{qvE}^3 V^2}\left[(\nabla\Delta\rho_{qvE})^2 - \frac{1}{c^2}\left(\frac{\partial}{\partial t}\Delta\rho_{qvE}\right)^2\right] \quad (13)$$

namely describes the geometry via the pressures that arise by the collisions between the virtual particles-antiparticles populating the vacuum corresponding to the **RS** processes. In this picture, the quantum potential of the vacuum (13) may be considered as the real origin of the quantum effects [16]. In particular, the non-local correlations characterizing the quantum domain can be seen as effects deriving from the general Bell length of the 3D quantum vacuum

$$L_{quantum} = \frac{c^2\hbar}{D\sqrt{\frac{V}{n}\left[-(\nabla\Delta\rho_{qvE})^2 + \frac{1}{c^2}\left(\frac{\partial}{\partial t}\Delta\rho_{qvE}\right)^2 - \Delta\rho_{qvE}\left(\nabla^2\Delta\rho_{qvE} - \frac{1}{c^2}\frac{\partial^2}{\partial t^2}\Delta\rho_{qvE}\right)\right]}}$$

(14).

The quantum length (92) is the ultimate parameter that indicates that, at a fundamental level, the 3D quantum vacuum defined by **RS** processes of creation/annihilation of virtual particles-antiparticles organized in Bose ensembles and corresponding to fluctuations of the quantum vacuum energy density, is a non-local and timeless manifold.

The maximum value of the Bell length of the 3D quantum vacuum (14), which implies the maximum de-localization of a quantum system, is 1, which means

$$\frac{2\Delta\rho_{qvE} V^{1/2}}{n^{1/2}\sqrt{\left[-(\nabla\Delta\rho_{qvE})^2 + \frac{1}{c^2}\left(\frac{\partial}{\partial t}\Delta\rho_{qvE}\right)^2 - \Delta\rho_{qvE}\left(\nabla^2\Delta\rho_{qvE} - \frac{1}{c^2}\frac{\partial^2}{\partial t^2}\Delta\rho_{qvE}\right)\right]}} = 1$$

(15).

Hence one obtains the following simple relation satisfied by the number of virtual particles-antiparticles of the **RS** processes of the 3D quantum vacuum in the condition of maximum entanglement, of the maximum grade of non-locality and de-localization in a quantum system having the mass $m = \frac{\Delta\rho_{qvE} V}{c^2 n}$ produced by the fluctuations of the quantum vacuum energy density corresponding to the same **RS** process:

$$n^{1/2} = \frac{2\Delta\rho_{qvE} V^{1/2}}{\sqrt{\left[-(\nabla\Delta\rho_{qvE})^2 + \frac{1}{c^2}\left(\frac{\partial}{\partial t}\Delta\rho_{qvE}\right)^2 - \Delta\rho_{qvE}\left(\nabla^2\Delta\rho_{qvE} - \frac{1}{c^2}\frac{\partial^2}{\partial t^2}\Delta\rho_{qvE}\right)\right]}}$$

(16).

The other important feature of the 3D quantum vacuum of AR is its action as a super-fluid medium. Taking account of Sbitnev's results [17-20], in which the physical vacuum is described as a super-fluid medium, containing pairs of particles-antiparticles which give rise a Bose-Einstein condensate, in **AR** it is assumed that, in the presence of ordinary baryonic matter, the

3D quantum vacuum physically acts as a super-fluid medium, which consists of an enormous amount of **RS** processes of creation/annihilation of particles-antiparticles with opposite orientations of spins. As a consequence of the motion of the virtual particles corresponding to the elementary fluctuations of the energy density, space is filled with virtual radiation with frequency

$$\omega = \frac{2\Delta\rho_{qvE} V}{\hbar n} \quad (17).$$

In the light of equation (17), each elementary fluctuation of the energy density of space in a given volume produces an oscillation of space at a peculiar frequency. This means that each material object given by mass (2) corresponds to oscillations of the 3D quantum vacuum given by equation (17).

The total effect of the motion of the virtual particles produced by the amount of **RS** processes characterizing a given region – in correspondence to changes of the energy density of space – is to generate a dragging, pushing effect of the 3D quantum vacuum. In particular, one may describe the pushing effect of a region of volume V of space in a given point at a distance R from the centre of that volume by defining a velocity of the 3D quantum vacuum on the basis of equation

$$v_{qv} = \frac{2\Delta\rho_{qvE} V}{\hbar n} R \quad (18).$$

The frequency (18) may also be considered the origin of the electromagnetic effects of the 3D quantum vacuum. In fact, the electromagnetic field inside a cavity of perfectly reflecting can be seen as an expansion of infinite different modes of the fundamental 3D quantum vacuum where each mode corresponds to an independent oscillation defined by frequency (17) produced by a specific **RS** process of creation/annihilation of quanta in correspondence to elementary fluctuations of the 3D quantum vacuum [16]. In **AR** the inertial mass of an object emerges from the interacting fraction of an energy density characterizing electromagnetic properties of the 3D quantum vacuum which are determined by the frequencies associated to opportune **RS** processes of creation/annihilation corresponding to elementary fluctuations of the quantum vacuum energy density, according to equation

$$m_i = \left[4 \frac{V^4}{\hbar^2 \pi^2 n^3 c^5} \int \eta(\rho)(\rho_{pE} - \rho)^3 d\rho \right] \quad (19)$$

where c is the speed of light, $\eta(\omega)$ is the spectral factor, interacting with the body, which physically measures the relative strength of the interaction between the oscillations produced by the motions of the virtual particles of the **RS** processes and the massive object, interaction which acts to oppose the acceleration. As a consequence, the explanation of the weak equivalence principle provided by Haisch, Rueda and Puthoff gets a new simple, suggestive and more unifying re-reading: here, the equivalence principle does not need to be independently postulated but derives directly as a consequence of the **RS** processes, and thus of the elementary fluctuations of the energy density, of the same 3D quantum vacuum.

Moreover, as shown in [2, 3], **AR** allows us to provide a unifying approach suggesting that the real explanation for the dark energy lies in the fluctuations of the 3D quantum vacuum energy density. This approach introduces the interesting perspective to interpret the curvature of space-time associated with a dark energy density as a consequence of more fundamental changes of the 3D energy density of space, through a quantized metric, characterizing the underlying microscopic geometry of the 3D quantum vacuum, expressed by relation

$$d\hat{s}^2 = \hat{g}_{\mu\nu} dx^\mu dx^\nu \quad (20)$$

where here the (quantum operators) coefficients of the metric are defined (in polar coordinates) as

$$\hat{g}_{00} = -1 + \hat{h}_{00}, \quad \hat{g}_{11} = 1 + \hat{h}_{11}, \quad \hat{g}_{22} = r^2(1+\hat{h}_{22}), \quad \hat{g}_{33} = r^2 \sin^2\vartheta(1+\hat{h}_{33}),$$
$$\hat{g}_{\mu\nu} = \hat{h}_{\mu\nu} \text{ for } \mu \neq \nu \quad (21)$$

and

$$\langle \hat{h}_{\mu\nu} \rangle = 0 \text{ except } \langle \hat{h}_{00} \rangle = \frac{8\pi G}{3}\left(\frac{\Delta\rho_{qvE}}{c^2} + \frac{35Gc^2}{2\pi\hbar^4 V}\left(\frac{V}{c^2}\Delta\rho_{qvE}^{DE}\right)^6\right) r^2 \text{ and}$$

$$\langle \hat{h}_{11} \rangle = \frac{8\pi G}{3}\left(-\frac{\Delta\rho_{qvE}}{2c^2} + \frac{35Gc^2}{2\pi\hbar^4 V}\left(\frac{V}{c^2}\Delta\rho_{qvE}^{DE}\right)^6\right) r^2 \quad (22).$$

In this picture, dark energy represents itself structured energy of space on the basis of equation

$$\rho_{DE} \cong \frac{35Gc^2}{2\pi\hbar^4 V}\left(\frac{V}{c^2}\Delta\rho_{qvE}^{DE}\right)^6 \quad (23)$$

and the variable energy density of space (producing dark energy) acts as a two-point correlation function according to relation

$$\frac{c^4}{4\pi\hbar^4}\left(\frac{V}{c^2}\Delta\rho_{qvE}^{DE}\right)^6 \cong \int_0^\infty C(s)s\,ds \quad (24).$$

In synthesis, in the approach of **AR**, a 3D quantum vacuum consisting of an enormous amount of **RS** processes of creation/annihilation of particles-antiparticles with opposite orientations of spins and acting as a super-fluid medium is the fundamental background which determines a unifying view of gravity, electromagnetic fields and quantum behaviour of matter as different aspects of the same fluctuations of the quantum vacuum energy density.

4. Advanced Relativity combines entropy and syntropy

As regards the understanding of the nature of evolution, the most popular idea is that there are two main trends: one is thermodynamic concerning matter, evolving downwards, i.e. towards higher entropy, and one concerns life and consciousness, evolving upwards, i.e. towards higher organization. The main differences that characterize these two trends of the evolution are that, in contrast to physical systems in which the main trend is approaching thermodynamic equilibrium, living organisms are organized as hierarchical systems of biological functions acting against decay. As a consequence, a fundamental difference exists between physical order created by physical self-organization and biological organization. About this second trend in the evolution, the Hungarian scientist Grandpierre has shown that the increase of the organizational state, and therefore the consequent decrease of entropy, of the living organisms, is determined by the biological electromagnetic radiation, by biological photons [21, 22]. He computed the entropy content of the human being tied to *in vivo* radiation, showing that it is larger than the standard physical entropy of the materials of the living organism. In Grandpierre's approach, a living organism can be compared to a football team, in the sense that it is *organized* in a way that all the players are endowed with special tasks to perform certain functions (that are additional to their space-temporal coordinates), and these goal-oriented (teleological) functional activities are organized to achieve the ultimate aim of living and flourishing [23]. Here, in order to build an exact theoretical biology, the fundamental difficulty lies in resolving the incompatibility between deterministic physical equations and genuine biological teleology expressing an end-point "selection", goal-orientation and purpose. In order to approach this issue, Grandpierre recently proposed to generalize the least action principle, one of the most powerful tools of physics, so that it becomes compatible with life's ultimate aim of flourishing [24], thus allowing that living organisms select the endpoint of their biological processes to be compatible with their biological aims. Fundamentally, the biological principle of greatest action expresses the fact that all living organisms strive to maximize action, actively maintaining their states far away from thermodynamic equilibrium as long as possible.

In **AR** the two main trends of the evolution (of matter and life respectively) can be considered parts of a one universal process. In **AR** entropy is valid only for material objects. Space in which material objects exist has no entropy; it does not follow second law of thermodynamics. Matter is entropic state of primordial energy of space which itself is syntropic. In living organisms the syntropic energy of space is the physical basis for negative entropy (negentropy) of living organism. Evolution of life is a continuous process of decreasing of entropy which converges to the non-entropy state of the fundamental 3D quantum vacuum. The evolution of life on Earth is part of a universal process which runs in the entire universe and is developing towards the syntropic energy of space. Matter has property of "self-organization" because it exists in space which is syntropic. Physical homogeneity of the universe also implies biological homogeneity. In the entire universe matter has a tendency to develop into intelligent conscious beings.

The idea of an eternal universe ruled by syntropy is not familiar with the cosmological models prevailing today, although all experimental data are on the side of such an eternal universe. The universe is eternally NOW and space plays an active role into evolution of life; it is sending information into 3D quantum vacuum which generates living organisms. Evolution of life cannot be imagined without space and 3D quantum vacuum as originators of life. All over the universe matter has tendency to develop into intelligent and conscious organisms whose evolution tends to rediscover the fundamental 3D quantum vacuum itself.

5. In Advanced Relativity the model of consciousness which is the origin of the observer is n-dimensional Hilbert space

AR is based on experimental data, namely NASA results confirms universal space is "flat". This means universal space has Euclidean shape and is not "finite", but it is infinite. From mathematics we know that "infinite distance + 1000km" is still infinite distance. Infinity in mathematics and physics is not a metrical term. Infinity of universal space we cannot fully comprehend by the rational mind, we can experience it in meditation. For millenniums spiritual researchers have experience of "infinite spaciousness" when meditating on the subject of human real nature - consciousness. Buddhists use term "Emptiness", "Nothingness", "Shunyata, Taoist use term "Tao". **AR** is enlarging reductionist approach of physics that only the phenomena exist which can be measured with an integral approach which confirms that also what can be experienced (without being measured) is real.

Consciousness is widely considered to be the greatest challenge for modern science. Consciousness may exist not in, but beyond, the experienced dimension, in the deep dimension of the cosmos. The idea that consciousness belongs to another, deeper dimension of reality has been frequently voiced not only by poets but by scientists. Erwin Schrödinger said, "The overall number of minds is just one. . . . In truth, there is only one mind.". In his later years psychologist Carl Jung came to a similar conclusion, claiming that the psyche is not a product of the brain; it is part of the generative, creative principle of the cosmos — of the *unus mundus*. Consciousness which is the origin of the observer is real and can be described as n-dimensional Hilbert space (where n is cardinal number of natural numbers). In physics we observe photon in a 3D material world. When frequency of the 3D photon is increasing to the infinite value, it is also entering higher dimensions of Hilbert spaces. Finally in n-dimensional Hilbert space energy of the photons is transformed into energy of consciousness. In a human being the neural networks of the brain resonate with the information associated with the n-dimensional Hilbert space. The human brain translates the information carried in that background in a holographically-distributed form into linear signals that affect the functioning of the brain's neural networks. In principle, the human brain is informed by the totality of the information in the deep-dimension of the n-dimensional Hilbert space.

In **AR** photon is the wave of universal space corresponding to the fundamental "3D quantum vacuum" or "3D physical vacuum". In quantum mechanics energy of the photon is:

$E = n \cdot h \cdot v$ (25),

where h is the Planck constant, v is the photon frequency and n is an integer number (1,2,3...). Consciousness can be described as the photon which has infinite frequency and exists in n-dimensional Hilbert space:

$$\lim_{v \to \infty} n \cdot h \cdot v = consciousness$$ (26).

In the formula (26) integer n represents the dimensionality of Hilbert spaces. When vibration v becomes infinite, integer n becomes cardinal number of natural numbers. Consciousness is the vibration of n-dimensional Hilbert space whose limit is tending to the infinite value of frequency and zero value of the wave length λ. Out of that it follows velocity v of consciousness is zero:

$v = v \cdot \lambda = 0$ (27).

This mathematical model cannot be considered as a real picture of the consciousness; it only indicates its real nature. Consciousness is beyond logic and so beyond mathematics which can only help us to build an approximate model. Mathematics cannot explain the physical reality of the universe; it can only describe it with its limitations.

The idea that consciousness can be associated with photons of infinite frequency is supported by recent experiments, which show that a bio-photon coherent system constitutes an ultra rapid communication system not only in the brain but also functioning in the whole organism, explaining the amazing concerted actions of complex living organisms. According to these experiments, bio-photonic and bio-electronic activities are not independent biological processes in the nervous system, and their synergistic action play a significant role in neural signal processes [25-27].

Consciousness does not move in space, consciousness is the fundamental vibration of space. When vibration of consciousness is getting lower it transforms into the light. Consciousness acts via bio-photons on the microtubules of the brain [1]. Information flow from the consciousness to the three-dimensional living matter which is composing living organisms is running from the n-dimensional Hilbert space via less dimensional Hilbert spaces until is arriving to the three-dimensional matter. Each "information jump" on the space with lower number is accompanied with the lower energy of the "photon" accordingly to equation (25) in which we can see the Planck constant is the constant which bridges different Hilbert spaces.

Our standard scientific view that the vibration can exist only in a 3D reality is now upgraded with the understanding that also higher dimensions are existent and that they have their own vibrations. Planck constant is the constant connecting these n-dimensional vibrations.

Photon observed in a 3D world is connected with the consciousness via Hilbert spaces. That is why bio-photons discovered by Russian physicists Alexander Gurwitsch Gurjijef and German physicist Albert Fritz Popp are of the immense importance. Bio-photons are an "information bridge" between atom level and high dimensional Hilbert spaces. Information which flows from the higher Hilbert spaces to the lower Hilbert spaces is designing not only functioning of the single living organism, but also entire evolution of life which is developing in the entire universe.

From the formalism (25) we can see that energy of primordial consciousness in n-dimensional Hilbert space has infinite energy. This "infinite energy" is spiritual energy, the energy of the pure Spirit-Consciousness. 3D world exists within this poll of infinite Spirit-Consciousness energy. With descending in lower Hilbert spaces this spiritual energy is becoming more and more "structured" and "dense" and appears in a 3D world as the photon. Recent discoveries confirm matter is made out of photons [28], which mean that matter is consciousness in its most dense form.

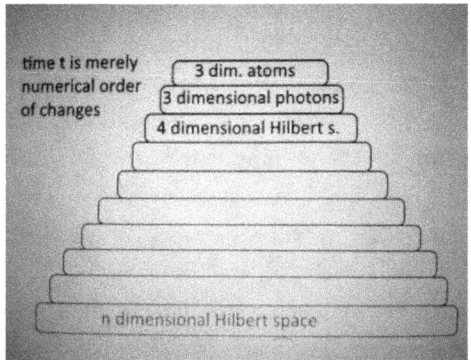

Figure 1: AR structure of the universe where photon is the wave of a 3D quantum vacuum and time has merely mathematical existence

6. N-dimensional Hilbert spaces, consciousness and De Broglie-Bohm pilot wave model in 3D quantum vacuum

Advanced Relativity (AR) can be also considered as a development of De Broglie-Bohm pilot wave model. Every elementary particle in a 3D universal physical space has its pilot waves in higher dimensional Hilbert spaces. Consciousness is the higher Hilbert space and is "guiding" via lower dimensional spaces 3D elementary particles.

The idea that consciousness can be related to Hilbert spaces is not new in the current research. For example, in [29, 30] Martin and Carminati studied the individual unconscious and consciousness as quantum systems, i.e., as vectors of a Hilbert space. In such a frame they studied the phenomenon of consciousness and especially the awareness of unconscious components. Writing down the state of the unconscious as $|U\rangle$ and the state of consciousness as $|C\rangle$, they introduced another state of the unconscious $|I\rangle$ which is the insight or pre-consciousness. By building a model of quantum entanglement between those three states they applied it to the awareness of unconscious components. In a similar way, starting from Jung's psychoanalytical perspective on mind and building on cognitive systems theory based on the mathematics of a Hilbert space, van den Hooff recently developed a formalism in which mental states, conscious as well as unconscious, are described by state vectors in a Hilbert space in a picture where the mental process of metabolization of blurred, semiconscious, or dreamlike mental states into clear ideas and unambiguous cognitive states can be formalized using eigenvalue equations [31].

According to our model everything in the world is ruled by a wave of primordial energy of n-dimensional Hilbert space. Consciousness acts as a wave belonging to a universal n-dimensional Hilbert space and governs the behaviour of particles and material objects via lower dimensional spaces.

Information theory indicates that there is a multi-dimensional connection between radiant energy and information signals "carried" by that energy, and that somewhat radiant energy and information are aspects of the same thing. It was Claude Shannon (a cousin of Thomas Edison) who in 1946 first discovered that thermodynamic equations could be used to relate the flow of thermal (radiant heat) energy to the flow of information signals encoded in any medium.

The branch of mathematical physics pioneered by Shannon and the science of cybernetics developed by Norbert Wiener at MIT led to methods of modulating information signals in a wide range of radiant energy spectrums from radio waves to visible light waves. Using the Fourier integral to analyze brain waves, Wiener described how frequency centers within the brain tend to attract one another, resonate and "self tune" in the frequency domain: "We thus see that a non-linear interaction causing the attraction of frequency can generate a self-organizing system, as it does in the case of the brain waves we have discussed...This possibility of self-organization is by no means limited to the very low frequency of these two phenomena. Consider self-organizing systems at the frequency level, say, of infrared light or radar spectra" [32].

Two spectral domains, spacetime and frequency are of essential importance in this field of information theory and signal communication. On the first page of the his standard textbook on electronic network information theory, Kuo states: "In describing signals, we use the two universal languages of electrical engineering – *time* and *frequency*. Strictly speaking, a signal is a function of time. However, the signal can be described equally well in terms of *spectral* or *frequency* information. As between any two languages, such as French and German, translation is needed to render information given in one language comprehensible in the other. Between time and frequency, the translation is effected by the *Fourier series* and the *Fourier integral*." [33].

In the light of these considerations and results, in our AR approach to consciousness, **the** Fourier transforms, which decompose a function of time (a *signal*) into the frequencies that make it up, namely transform time-space signals out of the space-time domain into a conceptually mysterious frequency domain, in which information is no longer measured by time or distance but by frequency and signal strength (amplitude), play a fundamental role. The Fourier integral transform of a continuous time function into the frequency domain is defined by relation

$$f(t) = \int_{-\infty}^{+\infty} g(v) e^{2\pi i v t} dv \quad (28)$$

which implies that energy radiant signals and information signals are two different aspects of one and the same thing, namely energy signals with information content, but potentially existing in either, or both simultaneously, of two very different dimensions, one with a spacetime component, the other a spaceless atemporal frequency dimension.

The mathematics discovered by Fourier operate on vibrational frequency information of every range. Accordingly they must be involved in information processing of thought and communication at all levels and frequencies, between and among humans, animals, tectonic plates, and perhaps at even greater macro and micro scales, with frequencies corresponding to waves ranging from the physical size of the expanding universe (calculated to be currently about 47 billion light years in diameter) down to the Plank length. Suggestive scenarios might derive from these theories.

In particular, existence of the frequency domain and Bell's theorem would explain non-locality in quantum mechanics, synchronicity, and morphic resonance, and perhaps telepathy and precognition.

In the AR approach, the crucial point is that Fourier transformations can describe transformation of waves from n-dimensional to n-1 dimensional Hilbert space. In order to show this, let us consider the fundamental equations of evolution of the RS processes occurring in the 3D quantum vacuum (5)-(12). By applying the Fourier Transform, equations (6) and (7) concerning the evolution of creation events and destruction events become respectively

$$\left(-\hbar^2 \partial^\mu \partial_\mu + \frac{V^2}{c^2}(\Delta \rho_{qvE})^2\right)\int_{-\infty}^{+\infty} g_{Q,i}(v)e^{2\pi i v t} = 0 \quad (29)$$

and

$$\left(\hbar^2 \partial^\mu \partial_\mu - \frac{V^2}{c^2}(\Delta \rho_{qvE})^2\right)\int_{-\infty}^{+\infty} h_{Q,i}(v)e^{2\pi i v t} = 0 \quad (30),$$

where

$$\psi_{Q,i}(t) = \int_{-\infty}^{+\infty} g_{Q,i}(v)e^{2\pi i v t} dv \quad (31)$$

and

$$\phi_{Q,i}(t) = \int_{-\infty}^{+\infty} h_{Q,i}(v)e^{2\pi i v t} dv \quad (32),$$

$g_{Q,i}(v)$ and $h_{Q,i}(v)$ being the frequency modes characterizing the creation and destruction processes respectively. In this way, in the quantum Hamilton-Jacobi equations (10)-(11), the quantum potential of the vacuum which the guides the occurring of the processes of creation or annihilation in space and makes the 3D quantum vacuum a fundamentally non-local manifold, assumes the following form

$$Q_{Q,i} = \frac{\hbar^2 c^2}{V^2 (\Delta \rho_{qvE})^2} \left(\frac{\left(\nabla^2 - \frac{1}{c^2}\frac{\partial^2}{\partial t^2}\right)\left|\int_{-\infty}^{+\infty} g_{Q,i}(v) e^{2\pi i v t}\right|}{\left|\int_{-\infty}^{+\infty} g_{Q,i}(v) e^{2\pi i v t}\right|} - \frac{\left(\nabla^2 - \frac{1}{c^2}\frac{\partial^2}{\partial t^2}\right)\left|\int_{-\infty}^{+\infty} h_{Q,i}(v) e^{2\pi i v t}\right|}{\left|\int_{-\infty}^{+\infty} h_{Q,i}(v) e^{2\pi i v t}\right|} \right) \quad (33).$$

According to equation (33), one can say that the non-local action of the quantum potential to guide the occurring of the processes of creation or annihilation in space is determined by the frequency modes which appear in the wave of the quantum vacuum. Consciousness is linked with the frequency modes $g_{Q,i}(v)$ and $h_{Q,i}(v)$ characterizing the creation and destruction events of quanta in the 3D quantum vacuum. The evolution of the elementary particles we experience occurs in a 3D space but the waves which guides the creation and destruction events determining their behaviour are associated to frequency modes that, in line of principle are infinite and, therefore, "live" in higher dimensional Hilbert space. According to equations (31) and (32), which relate the waves associated with the creation and destruction events to their corresponding frequency modes, in virtue of the Fourier transformations, the frequency modes characterizing the creation and destruction events of quanta in the 3D quantum vacuum belong to higher dimensional Hilbert spaces. These frequency modes are, in line of principle, infinite and it is according to Fourier transforms (31) and (32) that they make possible the fact that the behaviour of elementary particles is ruled in a 3D background. This means, in other words, that elementary particles we experience and live in a 3D space are guided by waves which ultimately originate from higher dimensional spaces, just associated with the frequency modes appearing in their Fourier decompositions. The approach here introduced, based on equations (29)-(33), implies that consciousness can be defined as the higher Hilbert space which is "guiding" via lower dimensional spaces 3D elementary particles.

7. Conclusions

Vital force of physics is constant re-examination about the validity fundamental principles of physics. The principle of space-time as a fundamental arena of the universe has no experimental verification and after 100 years of use did not contribute to the advance of physics. In equations of SR and GR and in physics in general symbol for time t means duration of changes running in space. Experimental data show that space and time are two different principles. Advanced Relativity is taking this in account. The result is renewed Relativity Theory which has turned from mathematical theory to physical theory where variable energy of space is the driving force of physical world on the micro and macro scale. Consciousness which is the origin of the observer is fundamental principle of the universe which is acting on matter via n-dimensional Hilbert spaces.

References:

1. Amrit Sorli, "On the Origin of the Observer", *American Journal of Modern Physics*, Vol. 3, No. 4, 2014, pp. 173-177. doi: 10.11648/j.ajmp.20140304.14.

2. D. Fiscaletti and A. Sorli, "Space-time curvature of general relativity and energy density of a three-dimensional quantum vacuum", *Annales UMCS Sectio AAA: Physics*, 69, 55-81 (2015). DOI: 10.1515/physica-2015-0004.

3. D. Fiscaletti and A. Sorli, "Dynamic Quantum Vacuum and Relativity", *Annales UMCS Sectio AAA*, accepted for publication in August 2016.

4. A. Sorli, M. Mageshwaran and D. Fiscaletti, "Energy - Mass - Gravity Theory", *American Journal of Modern Physics*. Special Issue: Insufficiency of Big Bang Cosmology, Vol. 5, No. 4-1, 2016, pp. 20-26. doi: 10.11648/j.ajmp.s.2016050401.

5. D. Fiscaletti and A. Sorli, "Perspectives of the Numerical Order of Material Changes in Timeless Approaches in Physics", Foundations of Physics, 45, 2, 105-133 (2015).

6. A. Sorli, V. Koroli, A. Nistreanu and D. Fiscaletti, "Cosmology of Einstein's NOW", *American Journal of Modern Physics*. Special Issue: Insufficiency of Big Bang Cosmology. Vol. 5, No. 4-1, 2016, pp. 1-5. doi: 10.11648/j.ajmp.s.2016050401.11.

7. A. Sorli, "Relative velocity of material change in a 3D quantum vacuum", Journal of Advanced Physics, Vol. 1, No. 1, pp. 110-112 (2012) DOI: 10.1166/jap.2013.1087.

8. D. Fiscaletti and A. Sorli, "About a new suggested interpretation of special theory of relativity within a three-dimensional Euclid space", *Annales UMCS, Sectio AAA: PHYSICA*, Volume 68, Issue 1, Pages 39–62, ISSN

(Print) 0137-6861, DOI: 10.2478/v10246-012-0018-1, March 2014.
9. NASA, http://map.gsfc.nasa.gov/universe/uni_shape.html (2013).
10. C. Rovelli, *Physics World* **7**, 11, 1-5 (2003).
11. C. Rovelli, "Loop quantum gravity", <http://relativity.livingreviews.org/Articles/lrr-2008-5/> (2008).
12. C. Rovelli, "A new look at loop quantum gravity", <arXiv:1004.1780v1 [gr-qc]> (2010).
13. L. Chiatti, "The transaction as a quantum concept", in *Space-time geometry and quantum events*, I. Licata ed., pp. 11-44 (Nova Science Publishers, New York, 2014); e-print arXiv.org/pdf/1204.6636 (2012).
14. I. Licata, "Transaction and non-locality in quantum field theory", *European Physical Journal Web of Conferences* (2013).
15. D. Fiscaletti and A. Sorli, "Perspectives about quantum mechanics in a model of a three-dimensional quantum vacuum where time is a mathematical dimension", *SOP Transactions on Theoretical Physics* **1**, 3, 11-38 (2014).
16. D. Fiscaletti and A. Sorli, "About a three-dimensional quantum vacuum as the ultimate origin of gravity, electromagnetic field, dark energy … and quantum behaviour", Ukrainian Journal of Physics, 61, 5, pp 413-431 (2016).
17. V. Sbitnev, "From the Newton's laws to motion of the fluid and superfluid vacuum: vortex tubes, rings, and others", arXiv:1403.3900v2 [physics.flu-dyn] (2014).
18. V. Sbitnev, *Modern Physics Letters A* **30**, 35, 1550184 (2015); e-print arXiv:1507.03519v1 [physics.gen-ph].

19. V.I. Sbitnev, "Physical vacuum is a special superfluid medium", in *Selected Topics in Applications of Quantum Mechanics*, M.R. Pahlavani ed., pp. 345-373, InTech, Rijeka (2015).

20. V.I. Sbitnev, "Navier-stokes equation describes the movement of a special superfluid medium", *Foundations of Physics*, in press, 2015; e-print http://arxiv.org/abs/1504.07497.

21. A. Grandpierre, "Entropy and information of human organisms and the nature of life", *Frontier Perspectives*, Vol. 13, No. 2, 16-21 (2004).

22. A. Grandpierre, "Order, structure, physical organization and biological organization", *Frontier Perspectives*, 14, No. 1, 6-13 (2005).

23. A. Grandpierre and M. Kafatos, "Genuine Biological Autonomy: How can the Spooky Finger of Mind Play on the Physical Keyboard of the Brain?", in: Hanna, P. (ed.), *An Anthology of Philosophical Studies*, 2013; Vol. 7, pp. 83-98.

24. A. Grandpierre, "Biological Extension of the Action Principle: Endpoint Determination beyond the Quantum Level and the Ultimate Physical Roots of Consciousness", NeuroQuantology, 5, pp. 346-362 (2007).

25. B. Dotta and M.A. Persinger, "Increased photon emissions from the right but not the left hemisphere while imagining white light in the dark: The potential connection between consciousness and cerebral light", JCER2, 1463–1473 (2011).

26. B.T. Dotta, K.S. Saroka and M.A. Persinger, "Increased photon emission from the head while imagining light in the dark is correlated with changes in

electroencephalographic power: Support for Bókkon's Biophoton Hypothesis", Neurosci. Lett., 513, 151–154.10.1016/j.neulet.2012.02.021 (2012).

27. Y. Sun, C. Wang and J. Dai, "Biophotons as neural communication signals demonstrated by in situ biophoton autography", Photochem. Photobiol. Sci., 9, 315–322.10.1039/b9pp00125e (2010).

28. O. Firstenberg, T. Peyronel, Qi-Yu Liang, A.V. Gorshkov, M.D. Lukin and V. Vuletić, "Attractive photons in a quantum nonlinear medium", *Nature* 502, 71–75 (03 October 2013) doi:10.1038/nature12512.

29. F. Martin and G. Galli Carminati, "Synchronicity, Quantum Mechanics, and Psyche", talk given at the Conference on "Wolfgang Pauli's Philosophical Ideas and Contemporary Science", May 20-25, 2007, Monte Verità, Ascona, Switzerland; published in *Recasting Reality*, pp. 227- 243, Springer-Verlag, 2009.

30. F. Martin, F. Carminati and G. Galli Carminati, "Synchronicity, Quantum Information and the Psyche", The Journal of Cosmology, 3: 580-589 (2009).

31. H. Van den Hooff, Mind and Matter, Vol. 11, n. 1, pp. 45-60 (2013).

32. N. Wiener, *Cybernetics or Control and Communication in the Animal and the Machine*, (Massachusetts: MIT Press, 1948), 200.

33. F.F. Kuo, *Network Analysis and Synthesis*, (New York: John Wiley & Sons, 1962), 1.

Advanced Relativity: Multidimensionality of Consciousness and Mind, Origin of Life, PSI Phenomena

Abstract

Evolution of life has its informational basis in higher dimensional Hilbert spaces which also represent the origin of consciousness and mind. The idea of neurology that mind has its origin in neuronal activity of the brain, which is three-dimensional, is here improved with the model that the mind states exist in higher dimensional Hilbert spaces. Mind is multidimensional and has the ability of PSI phenomena.

Key words: consciousness, mind, origin of life, PSI phenomena.

1. Introduction

All thoughts, theories and mathematics that the human mind can produce can be observed by the observer, which in Advanced Relativity has its origin in n-dimensional Hilbert space (Sorli et al., 2016). This means that the observer has the highest degree of complexity, which is then followed by mind in lower dimensional Hilbert spaces.

Advanced Relativity (AR) can be considered as an improvement and completion of Einstein's Relativity with the help of quantum field theory (in particular, as regards the concept of creation/annihilation of elementary quanta). In AR the four-dimensional space of general relativity derives from a more fundamental three-dimensional quantum vacuum where time exists only as a measuring system of the numerical order of material changes. More precisely, the four dimensions of universal space can be seen as the effects, at an upper/explicate level, of a more fundamental three-dimensional timeless non-local quantum vacuum defined by reduction-state (RS) processes of creation/annihilation of particles/antiparticles (with opposite orientations of spins), corresponding to elementary fluctuations of the quantum vacuum energy density. The RS processes are described by a wave function at two components satisfying a time-symmetric extension of the Klein-Gordon quantum relativistic equation

$$\begin{pmatrix} H & 0 \\ 0 & -H \end{pmatrix} C = 0 \qquad (1)$$

where

$$H = \left(-\hbar^2 \partial^\mu \partial_\mu + \frac{V^2}{c^2} (\Delta \rho_{qvE})^2 \right) \quad (2)$$

and

$$\Delta \rho_{qvE} = \rho_{pE} - \rho_{qvE} \qquad (3),$$

$$\rho_{qvE} = \rho_{pE} - \frac{mc^2}{V} \qquad (4),$$

where m is the mass of the object, V is its volume,

$$\rho_{pE} = \frac{m_p \cdot c^2}{l_p^3} \quad (5)$$

is the Planck energy density defining the ground-state of the 3D quantum vacuum (where m_p is Planck mass, and l_p is Planck length). The wave functions characterizing the RS processes of creation and annihilation are respectively

$$\psi_{Q,i}(t) = \int_{-\infty}^{+\infty} g_{Q,i}(v) e^{2\pi i v t} dv \qquad (6)$$

and

$$\phi_{Q,i}(t) = \int_{-\infty}^{+\infty} h_{Q,i}(v) e^{2\pi i v t} dv \qquad (7),$$

$g_{Q,i}(v)$ and $h_{Q,i}(v)$ being the frequency modes characterizing the creation and destruction processes respectively. The occurring of the processes of creation/annihilation of quanta in space is determined by the frequency modes which appear in the wave function of the quantum vacuum through the quantum potential of the vacuum

$$Q_{Q,i} = \frac{\hbar^2 c^2}{V^2(\Delta\rho_{qvE})^2} \left(\frac{\left(\nabla^2 - \frac{1}{c^2}\frac{\partial^2}{\partial t^2}\right)\left|\int_{-\infty}^{+\infty} g_{Q,i}(v)e^{2\pi i v t}\right|}{\left|\int_{-\infty}^{+\infty} g_{Q,i}(v)e^{2\pi i v t}\right|} - \frac{\left(\nabla^2 - \frac{1}{c^2}\frac{\partial^2}{\partial t^2}\right)\left|\int_{-\infty}^{+\infty} h_{Q,i}(v)e^{2\pi i v t}\right|}{\left|\int_{-\infty}^{+\infty} h_{Q,i}(v)e^{2\pi i v t}\right|} \right) \quad (8)$$

which makes the 3D quantum vacuum a fundamentally non-local manifold.

In AR, the 3D quantum vacuum consisting of an enormous amount of RS processes of creation/annihilation of particles-antiparticles with opposite orientations of spins can provide a unifying view of gravity, electromagnetic fields and quantum behaviour of matter as different aspects of the same fluctuations of the quantum vacuum energy density. In particular, as a consequence of the evolution of RS processes, spacetime is filled with virtual radiation of frequency

$$\omega = \frac{2\Delta\rho_{qvE}V}{\hbar n} \quad (9).$$

In the light of equation (9), each elementary fluctuation of the quantum vacuum energy density in a given volume produces an oscillation of space at a peculiar frequency. This means that each material object corresponds to oscillations of the 3D quantum vacuum given by equation (9).

The frequency (9) may be considered the origin of the electromagnetic effects of the 3D quantum vacuum in the sense that the electromagnetic field inside a cavity of perfectly reflecting can be seen as an expansion of infinite different modes of the fundamental 3D quantum vacuum where each mode corresponds to an independent oscillation defined by frequency (9) produced by a specific RS process of creation/annihilation of quanta in correspondence to elementary fluctuations of the 3D quantum vacuum (Fiscaletti and Sorli, 2016). This means that the spectral energy density for the zero-point fluctuations characterizing the electromagnetic properties of the quantum vacuum is

$$\rho(\Delta\rho_{qvE}) = \frac{4(\Delta\rho_{qvE})^3 V^3}{\hbar^2 \pi^2 n^3 c^3} \quad (10).$$

By starting from equations (9) and (10), in AR the electric and magnetic fields are two different kinds of polarization of the 3D quantum vacuum produced by the frequencies of the radiation associated with the motion of the virtual particles produced in the RS processes, namely by the elementary fluctuations of the quantum vacuum energy density. In particular, in the SED regime, the quantum vacuum fluctuations are random plane waves summed over all possible modes with each mode having the zero-point energy $\hbar\omega/2$, and thus the electric and magnetic fields may be expressed as

$$\vec{E}^{zp}_\tau(\vec{r},t) = \sum_{\lambda=1}^{2} \int d^3k (\Delta\rho_{qvE} V / n\pi^2)^{1/2} \hat{\varepsilon}(\vec{k},\lambda) \cos\left[\vec{k}\cdot\vec{r} - \frac{2\Delta\rho_{qvE} V}{\hbar n} t - \theta(\vec{k},\lambda)\right]$$

(11),

$$\vec{B}^{zp}(\vec{r},t) = \sum_{\lambda=1}^{2} \int d^3k (\Delta\rho_{qvE} V / n\pi^2)^{1/2} [\hat{k}\times\hat{\varepsilon}(\vec{k},\lambda)] \cos\left[\vec{k}\cdot\vec{r} - \frac{2\Delta\rho_{qvE} V}{\hbar n} t - \theta(\vec{k},\lambda)\right]$$

(12) (Fiscaletti and Sorli, 2016).

According to relations (11) and (12), the electromagnetic radiations are expressed as expansions of plane waves, where the sum is over two polarization states, $\hat{\varepsilon}$ is a unit vector, $\vec{k}$ is the polarization vector such that $|\vec{k}| = \omega/c$ and $\theta(\vec{k},\lambda)$ is a random variable uniformly distributed in the interval $(0, 2\pi)$ and independently for each wave vector $\vec{k}$ and polarization index λ. The magnetic field can be seen as the polarization of space in $\hat{k} \times \hat{\varepsilon}(\vec{k},\lambda)$; the electric field is the polarization of space in $\hat{\varepsilon}(\vec{k},\lambda)$.

Photons defined by the fields (11) and (12) can be considered the fundamental entities which produce the storing of information. In AR, microtubules are getting information via bio-photons from 3D quantum vacuum, an idea that is supported by recent research suggesting that microtubules use a binary system for storing information (Craddock et al. 2016). Bio-photons have spin angular momentum $\pm\hbar$, where $\hbar$ is the reduced Planck constant and the $\pm$ sign is positive for right and negative for left circular polarizations. Left-handed bio-photons carry information of zero (0) and right-handed bio-photons carry information of (1). Each bio-photon carries thus one bite of information which can be zero (0) or one (1). By considering the helicity $H = \dfrac{\vec{\sigma}\cdot\vec{p}}{|\vec{\sigma}||\vec{p}|}$, where $\vec{\sigma}$ is the spin vector and $\vec{p}$ is the momentum, since under space inversion $\vec{\sigma}\cdot\vec{p}$ changes sign, the net helicity of bio-photons associated with the parity-conserving electromagnetic interactions must be zero and, therefore, right-handed and left-handed bio-photons are emitted or absorbed with equal amplitudes.

The idea that information and consciousness are spin-mediated is not new. A spin-mediated consciousness theory was proposed by biophysicist Huping Hu with his collaborator Maoxin Wu (2002, 2004 and 2008) and then reviewed by Sultan Tarlaci (2006). According to this theory, consciousness is intrinsically connected to the spin process, which provides the linchpin between mind and the brain, and emerges from the self-referential collapses of spin states and the unity of mind is achieved by entanglement of these mind-pixels. The starting epistemological foundation of this theory lies in the fact that spin is basic quantum bit ("qubit") for encoding information and, on the other hand, neural membranes and proteins are saturated with nuclear spin carrying nuclei and form the matrice of brain electrical activities.

On the other hand, Persinger's group measured significant increases in biophoton emissions along the right side with respect to the left side, when subjects imagined white light in a dark environment. These quantitative measurements as well as quantitative analysis strongly suggest that spin energies can accommodate the interactions between protons, electrons, and photons and the action potentials associated with intention, consciousness and entanglement (Persinger et al., 2013).

According to the approach of AR suggested in this paper, polarizations of biophotons are the fundamental elements originating consciousness through the following mechanism. A photon of the 3D quantum vacuum has a correspondent "pilot photon" in the 4D Hilbert space. Regarding the 3D photon, the 4D pilot photon, which is the vibration of the 4D Hilbert space, has higher density of information according the formula below:

$$C_k(n) = \binom{n}{k} = \frac{n!}{k!(n-k)!} = \frac{4!}{3!(4-3)!} = 4 \qquad (13).$$

In the light of equation (13), the 3D photon can manifest in the following four possible states of 4D pilot photon: $\psi(X_1, X_2, X_3)$, $\psi(X_1, X_2, X_4)$ $\psi(X_1, X_3, X_4)$, $\psi(X_2, X_3, X_4)$. This means that the 4D pilot photon contains 4 combinations of 3D photons, which implies it has 4 bites of information. These 4 bites can have according to the formula (13) 6 combinations, namely:

$$C_k(n) = \binom{n}{k} = \frac{n!}{k!(n-k)!} = \frac{4!}{2!(4-2)!} = 6 \qquad (14),$$

$(4_1 4_2), (4_1 4_3), (4_1 4_4), (4_2 4_3), (4_2 4_4), (4_3 4_4)$,

where n is 4 and k is 2, namely (0) and (1).

6 binary combinations (which are 6 bites of information) of the 4D pilot photon can be than transformed in the one bite information of 3D photon which can be (0) or (1). When the 3D photon is moving through the microtubule it can give 6 binary impulses which are guided by the 4D pilot photon. 3D photon is the last element in the "energy chain" of pilot photons in higher dimensional Hilbert spaces. 5D pilot wave has 10 binary combinations calculated by the formula (15):

$$C_k(n) = \binom{n}{k} = \frac{n!}{k!(n-k)!} = \frac{5!}{2!(5-2)!} = 10 \quad (15),$$

where n is the number of Hilbert spaces and $k = 2$.

10 binary combinations of the 5D pilot photon can be than transformed in the 6 bites of information of the 4D photon. 6D pilot wave has 15 combinations which can be transformed in the 10 bites of information of 5D pilot photon and so on.

3D - 1 bite
4D - 6 bites
5D - 10 bites = 6 + 4
6D - 15 bites = 10 + 5
7D - 21 bites = 15 + 6
8D - 28 bites = 21 + 7
9D - 36 bites = 28 + 8

We see that the binary combinations (and so bites of information) in 5D and higher dimensional Hilbert spaces are increasing by the formula:

$$n \geq 5 \rightarrow X_n = X_{n-1} + (n-1) \quad (16).$$

In the Advanced Relativity model, the density of information increases in higher dimensional Hilbert spaces. The density of information is correlated with intelligence, which means that consciousness as the n-dimensional Hilbert space is of infinite intelligence. Human senses are three-dimensional and can perceive only the three-dimensional aspect of energy-information, in the form of ordinary 3D space, 3D particles and 3D massive bodies.

3D quantum vacuum which gives origin to the physical space is syntropic type of energy (Sorli et al., 2016). The concept of syntropic phenomena intended as those processes responsible of the evolution of the biological complexity, was originally introduced by Fantappiè in 1941 in his unifying theory of physical and biological world and then was further developed by his collaborators Giuseppe and Salvatore Arcidiacono. In the approach of Fantappiè-Arcidiacono-Arcidiacono, life consists in a set of syntropic phenomena and cannot be only reduced to physical and chemical processes and therefore a source of information exists which supervenes chemical and physical processes with the aim of creation of biological systems. In the Advanced Relativity model, 3D photons, which are waves of 3D space, are the fundamental objects which enable living systems to lower the entropy of their environment (and thus provide the ultimate source of information which is responsible of creation of the evolution of biological complexity, in epistemological affinity with the Fantappiè-Arcidiacono-Arcidiacono view). Fritz-Albert Popp calls these photons in living organisms bio-photons (Cohen and Popp, 1997). The actual structure of bio-photons is not different from ordinary photons, and they are also 3D. Bio-photons of living organisms are actually the physical origin of "bio-energy", which in India is called "prana", and in China "QI". In AR the health of an organism is based on a good information flow between higher dimensional Hilbert spaces and the 3D material organism; and therefore wrong and partial imaginations of the mind are the main cause of the illness. This is why health is intrinsically related to the origin of the observer, which is consciousness. Consciousness is the primordial energy-information of the universe, which, acting via mind to the 3D level of the organism, provides psychophysical health. When mind is not supported by consciousness, it becomes hyperactive, resulting in too much electrical current in the nervous system and its damage. On the other hand, a mind delighted with consciousness is calm, has higher efficiency and is the basis of psychophysical health (Davidson et al., 2003; Rubia, 2009; Rosenkrantz et al., 2013). In this perspective health, stable peace and prosperity of the human society can be achieved with the introducing of individual research on the origin of the observer in the educational system worldwide. The observer is the common ground element of every human being regardless different cultural and religious environments, and as such represents the real binding force between different cultures and religions (Sorli, 2016).

In AR, by following a philosophy that characterizes quantum field theory, the notion of energy is extended into multidimensional Hilbert spaces. Energy can be defined in 3D Hilbert space, in 4D Hilbert space, in 5D Hilbert space, more in general in a nD Hilbert space where n is the cardinal number of natural numbers; nD Hilbert space is the fundamental background of consciousness, which acts in the human being as the observer. Information has its origin in energy, which exists in different Hilbert spaces; energy and information are the same reality, an idea which is today generally accepted: "Quantum mechanics is not the final stage in the science of physics. At the beginning of quantum mechanics between 1900 and 1950, everything was particles, but later, from 1950 to 1970, everything was fields. The view today – from 1970 on – is that everything is information" (Tarlaci, 2010). Moreover, in the model here proposed, the information of the higher dimensional Hilbert spaces ruling the phenomena of the universe can be associated with specific properties of the fundamental non-local 3D quantum vacuum.

Recent publications of O.J. Pike and others, O. Firstenberg and others are providing the basic understanding of photons being constitutive elements of the matter (Firstenberg et al., 2013; Pike et al., 2014). In AR photons are waves of 3D physical space (3D quantum vacuum), which means particles are different forms of space energy as already predicted by Ervin Schrödinger: "What we observe as material bodies and forces are nothing but shapes and variations in the structure of space". In this picture time as numerical order of phenomena is characteristic for 3D type of energy. Phenomena which exist in higher dimensional Hilbert spaces will be addressed in next chapter are immediate and so have not time. In nD reality we do not have phenomena any more. Consciousness is non-phenomenal, pure emptiness, "nothingness", shunyata, Nirvana.

The very nature of consciousness energy is infinite and as such unreachable for exact mathematical description with no epistemological gap. Advanced Relativity model has developed a mathematical picture for reality in which every element of the model corresponds exactly to one element in the real world, so there is no epistemological gap between the model and reality. Our research group works on the model with minimal epistemological gap for higher dimensional Hilbert spaces. Consciousness as non-phenomenal reality is remaining outside of this description. The true nature of consciousness research is not descriptive; it is experiential, individual research method (meditation) in which one discovers Nirvana: "Nirvana is a state of pure bliss and knowledge. ... It has nothing to do with the individual. The ego or its separation is an illusion" (Ervin Schrodinger).

2. Origin of Life

In AR life on the molecular level of 3D dimensionality has origin in nD consciousness. Evolution of life is encoded in 4D and more dimensional Hilbert spaces of higher information density and is communicated to the 3D ordinary life dimensionality via bio-photons. It is already confirmed bio-photons are transporting information between cells. This view is confirmed by discoveries in quantum bio-communication (Edmonds, 1994; Sun et al., 2010) as well as, above all, in a recent model of phonon guided biology, where the functional architecture and metabolic networks of cells in living organisms are associated with typical eigenfrequencies, and in particular electromagnetic eigenfrequencies have a relation to quantum coherence in life systems to brain function and consciousness (Meijer and Geesink, 2016). According to Meijer's and Geesink's approach, quantum wave effects acting on life systems are determined by a toroidal quantum field at typical scalars and related to Frölich-Bose Einstein condensates working in biological processes and present in phonons. From this perspective, reality may be perceived as a vast interlacing network of discrete fields of oscillations that interact with the vibrational fields of life systems, a communication process in which photons of various energy content and solitons play an essential role. Here, a relevant result is that geometrical information fluxes of photons and phonons converted to scalar standing waves in a 4D toroidal topology can be considered the basis for the origin of consciousness states. Moreover, research of Tang and Dai suggested that bio-photons also may play a potential role in neural signal transmission and processing, contributing to the understanding of the high functions of nervous system (Tang and Dai, 2014).

In this perspective, the evolution of life is an ongoing process running in the entire universe (Sorli, 2016). Essential organic molecules which build life have been found in the entire universal space (Kwok and Zhang, 2011). Cosmic energy forms a connected matrix within the entire cosmos and by means of structured matter can elicit life (Pereira, 2015a). Energy Exchange between particles and quantum vacuum is a relevant reason of the interdependence between the cosmic energy matrix and matter. Emergence of life within matter corresponds to the point of conversion from non-local cosmic energy to localized energy which could have been mediated by quantum processes across the developing proton gradients within organizing matter. Inorganic elemental matter could have triggered quantum based energy processing and the creation of the first living conscious forms (Reddy and Pereira, 2016).

According to AR physical homogeneity of the universe implies also biological homogeneity. The physical circumstances for development of life are the same in entire universal space. On the planets similar to the Earth life has developed. A planet with the similar circumstances for development of life as on the Earth was discovered recently (Jenkins et al., 2015). One cannot exclude that in the universe there exist other planets with the similar circumstances. Consciousness is shaping life via Hilbert spaces in entire universe. This also means that technology will never be able to create a "conscious computer" which would have higher cognitive abilities as friendship, empathy, compassion, which are in the domain of higher dimensional Hilbert spaces. Technology cannot reach beyond 3D ordinary reality.

3. PSI phenomena

The Advanced Relativity model implies that in higher dimensional Hilbert spaces (4D and more) the energy-information transfer – which can be ultimately associated with the action of specific properties of the fundamental non-local 3D quantum vacuum – is immediate. Telepathy between two or more human minds runs through higher dimensional Hilbert spaces where there is no time. A thought which is entering existence in a given human mind is immediately present in entire universe. In this sense two human minds are entangled and can communicate instantly via higher dimensional Hilbert spaces.

In the Advanced Relativity model, remote viewing (Tart et al., 1980) also happens via higher dimensional Hilbert spaces and is immediate. People with remote viewing capacity have developed "direct perception", which is beyond linear time as the "past-present-future" which is 3D activity of the human brain: "Traditionally, the way in which time is perceived, represented and estimated has been explained using a pacemaker-accumulator model that is not only straightforward, but also surprisingly powerful in explaining behavioural and biological data. However, recent advances have challenged this traditional view. It is now proposed that the brain represents time in a distributed manner and tells the time by detecting the coincidental activation of different neural populations" (Buhusi and Meck, 2005).

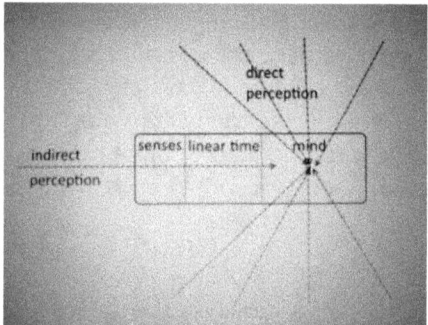

Figure 1: Indirect and direct perception

Some people have developed clairvoyance naturally from birth, as the capacity of the observer to see coming events on a 3D plane, which already are in the process of actualization in higher dimensional Hilbert spaces. What happened on 3D plane in an individual life and the life of entire human society has the beginning in higher dimensional Hilbert spaces, which also are the origin of the collective human mind. Social events on the large scale, as for example wars, have the source in "mental" tensions on the higher dimensional Hilbert spaces which then escalate in physical violence. Experiments confirm that continuous meditation of 2500 trained individuals has reduced crime in city DC Washington for 23% during the meditation period (Hagelin et al. 1999). Their meditation has brought peace and harmony in higher dimensional Hilbert spaces which has influenced the individual minds of entire population. Research of Radin and others confirms that directing intention toward a distant person is correlated with activation of that person's autonomic nervous system (Radin et al., 2008). Direct intention is directly transported via higher dimensional Hilbert spaces and is manifesting on 3D ordinary physical level.

The observer which is still experiencing world in the frame of linear psychological time, which has its origin in neuronal activity of the brain, has limitations in the understanding of telepathy, remote viewing, and clairvoyance. Albert Einstein used to say: "Time and space are modes by which we think and not conditions in which we live. Time has no independent existence apart from the order of events by which we measure it. There is something essential about the NOW which is just outside the realm of science. People like us, who believe in physics, know that the distinction between the past, present and future is only a stubbornly persistent illusion" (Sorli, 2016). Advanced Relativity model has confirmed that Einstein is right; time has only the mathematical existence and exists only in the 3D ordinary reality.

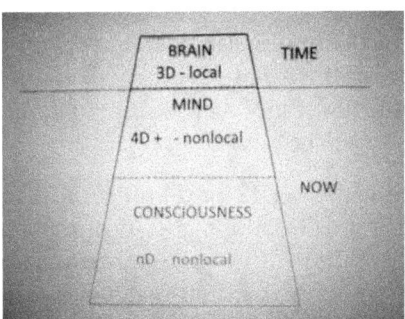

Figure 2: Brain, mind and consciousness in Advanced Relativity

4. Discussion

AR has developed a mathematical model for 3D ordinary background in which every element of the model corresponds exactly to one element in the real world, so there is no epistemological gap between the model and reality. The nature of 4 and more dimensional reality is such that the model of higher dimensional Hilbert spaces with higher information density is an approximate model with a certain epistemological gap. Consciousness as non-phenomenal reality is described by nD Hilbert space which does not reflect fully the very nature of consciousness which is not descriptive; it is experiential, individual research method (meditation) in which one discovers consciousness, also named emptiness, nothingness, Nirvana. Even Nobel Prize Laureate Erwin Schrödinger, who had a life-long interest in the Vedanta philosophy of Hinduism, which influenced his speculations at the close of his book *What is Life?* about the possibility that individual consciousness is only a manifestation of a unitary consciousness pervading the universe, recognized that: "Vedanta teaches that consciousness is singular, all happenings are played out in one universal consciousness and there is no multiplicity of selves. Nirvana is a state of pure bliss and knowledge. It has nothing to do with the individual. The ego or its separation is an illusion. Although I think that life may be the result of an accident, I do not think that of consciousness. Consciousness cannot be accounted for in physical terms. For consciousness is absolutely fundamental. It cannot be accounted for in terms of anything else" (Schrödinger, 1944).

In order to get more physical insight about Quantum Brain Dynamics it is necessary to carefully examine properties of a single protein microtubule of brain neurons from quantum-mechanical point of view (Enaki et al, 2016). This will give new possibilities for realization and manipulation of quantum information processes with the help of entanglement and non-local channels in the ensemble of microtubules of brain neurons (Satyajit et al., 2013).

From metaphysical point of view, a very interesting idea consists in the fact that while material universe is three-dimensional, the mathematical universe is multi-dimensional. Uniting theory and technologies for matter (quantum theory and physics, and nanotechnology), building blocks of living systems (biotechnology) and cognitive sciences (cogno) becomes necessary for developing new human capacities. Such initiatives by definition rely on converging existing state of the art trans-disciplinary research converging nano-bio-info-cogno disciplines, possibly under the more successful quantum perspective for understanding the world. Such confluence is imperative for mapping and analysing massive and complex brain data as, for example, measuring signals from nervous system, neuron to neuron interactions, conditions and repairs of sub-cellular structures for regenerative medicine, neuromorphic engineering, adaptation to change, collective and artistic human capabilities, broadening and transferring human perception, to name a few.

Because protein microtubules of brain neurons are quantum optical devices it is instructive to study the generation of squeezing and entanglement in microtubules to obtain more insight about nature of consciousness from quantum biophysical point of view (Satyajit et al., 2013) and role of consciousness in the Universe. In this situation non-local correlations takes place between different microtubules of brain neurons and therefore quantum information in the brain may be transmitted through non-local channels which in this paper are "pilot photons" of higher dimensional Hilbert spaces. The model of the mind and consciousness presented in this paper offers a deeper understanding of altered states of consciousness (ASC), induction of ASC by different physical methods: electrical stimulation (direct and alternating current), magnetic stimulation, sound (binaural) stimulation compared with placebo in healthy volunteers (Moldovanu and Vovc, 2013).

Several researches confirm that meditation (mindfulness) is reconnecting human mind with consciousness and so improving man's psychophysical health (Cojocaru and Moldovanu, 2013). Meditation is a technique which activates consciousness by "observing" (watching, witnessing) body-mind activities. The fundamental function of consciousness is the permanent awareness of the activity in lower dimensional Hilbert spaces (mind) and body which is 3D. This awareness has the power of body-mind healing, because it disintegrates mental forms which cause illness and replaces them with more adequate mental forms which are life supportive. Deepak Chopra's integrative medicine is the valuable practice of how meditation activates consciousness which is healing body and mind (de Jonng et al., 2016). Meditation is not only reorganizing higher dimensional Hilbert spaces (mind), it is also increasing cortical thickness of the brain (Chopra, 1990) which confirms the power of consciousness when acting on the 3D brain level.

Advanced Relativity model is fulfilling the request for deeper understanding of space-time metrics which will allow further development of 'Orch OR' Theory": "Hopefully, in the near future, with more experimental understanding of the space–time metric, Orch-OR would evolve to a complete deductive mathematical expression of consciousness—a dream that entire mankind is eagerly waiting to see" (Lazar et al., 2005). In AR space-time metrics is fully understood: time exists only as the numerical order of microtubule activity on the 3D physical reality. In 4D + Hilbert spaces where is always NOW information transfer is immediate. In AR the introduction of "imaginary time" in order to develop space-time metrics (proposed by Ghosh and others) is not necessary: "In addition, the fractal shape of entire brain architecture suggests that one clock is physically located inside another clock. Hence, we get information processing in an imaginary space. When we have both an imaginary space and an imaginary time, then we get a generalized hyper-dimensional space. Therefore, the discovery of resonance and wireless processing lead to a layered architecture of multiple space–time metric stacked one above another" (Ghosh et al., 2013). In 4D and more Hilbert spaces there is no more "space-time" metrics, we have only different layers of Hilbert spaces where is always NOW.

In the Advanced Relativity model, the mathematical definition of brain as 3D reality, mind as 4D + reality and consciousness as nD reality is unifying Platonic values of mathematics and Vedanta wisdom: "Orch OR provides a credible, testable model for how mental activity enters the physical world. I would take its optimism and turn it around: the mind-brain problem is indeed closer to being solved, not because quantum events give rise to mind but because these events indicate that an invisible agency (consciousness) is producing orderly, intelligent, information-infused activity at the very interface where spacetime emerges. The Platonic values of mathematics are undeniable, and once they are admitted into the picture, Vedanta would allow in every other Platonic value (truth, beauty, love). Then "nothing" – pure awareness without qualities – is the only viable explanation left standing for the origin of mind and reality itself" (Chopra, 2013). In this perspective meditation is the only valuable experiential research method to discover consciousness.

There is a certain correspondence with chakras in yoga and higher dimensional Hilbert spaces of AR:

3D - Muladhara
4D - Svadishthana
5D - Manipura
6D - Anahata
7D - Vishuddha
8D - Ajna

9D - Sahasrara (state of Nirvana)
nD – non-manifested consciousness (state of Parinirvana)

In this picture non-manifested (non phenomenal) consciousness is nD. Between 9D and nD there is a passage which in Buddhism represents the passing from Nirvana (9D) to Parinirvana (nD) which occurs upon the death of the body of someone who has attained Nirvana during his/her lifetime.

The thesis of biologist Bruce Lipton is that human mind empowered with consciousness can change genetic code (Lipton, 2007). According to the Advanced Relativity model it cannot be a-priory excluded that the person which has attained Nirvana (9D Hilbert space) has the power of the mind to change his genetic code. The 3D photon accompanied with 9D pilot photon can carry 36 bite of information which means it can hit 12 microtubule triplets and transmit them information. This information is then transferred via biochemical signals of polypeptides (MAPs) (Villasante et al. 1981), which then transport information to the 12 genome triplets. This is how information is transported from the "enlightened mind" to the genome.

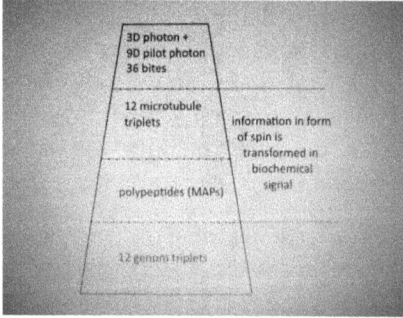

Figure 3: Information flow from the enlightened mind to the genome

Consciousness is the fundamental energy of the universe and has powers which seem that are uncomprehensible for common scientific mind which experiences life in the frame of linear time. The power of NOW (Tolle, 1997) has to be discovered firstly and then one attains the larger more adequate scientific view which allows him/her to see clearly the real nature of life.

5. Conclusions

Advanced Relativity is the unified model of the universe, life, mind, and consciousness where all phenomena are guided by consciousness via lower dimensional Hilbert spaces. In this model, time is the numerical order of changes and exists only in 3D reality. 4D and more dimensional Hilbert spaces realities are timeless (phenomena there do not have numerical order and are immediate) which gives new vision on the origin of life, PSI abilities and the timeless nature of consciousness itself.

References:

Arcidiacono G and Arcidiacono S. Entropia, sintropia, informazione, Di Renzo Editore, Rome, 1991.

Buhusi CV and Meck WH. What makes us tick? Functional and neural mechanisms of interval timing. Nature reviews 2005; 6: 755-764.

Chopra D. Quantum Healing, Bantam Books, New York, 1990.

Chopra D. Reality and Consciousness: A View From the East: Comment on "Consciousness in the Universe: A Review of the 'Orch OR' Theory" by Stuart Hameroff and Roger Penrose. Phys Life Rev **2013**; 11 (1): 81-100.

Cohen S and Popp FA. Biophoton emission of the human body. Journal of Photochemistry and Photobiology B: Biology 1997; 40(2): 187-189.

Cojocaru N and Moldovanu I. Altered states of consciousness induced by voluntary hyperventilation, Pilot study. Bul. Acad. Sci. of Moldova. Medical Sciences 2013; 2(38): 109-111.

Craddock TJA, Tuszynski JA and Hameroff S. Cytoskeletal Signaling: Is Memory Encoded in Microtubule Lattices by CaMKII Phosphorylation?. PLoS Comput Biol 2012; 8(3): e1002421.

Davidson R et al. Alterations in brain and immune function produced by mindfulness meditation. Psychosomatic medicine 2003; 65(4): 564-570.

de Jong M et al. Effects of Mindfulness-Based Cognitive Therapy on Body Awareness in Patients with Chronic Pain and Comorbid Depression. Front Psychol. 2016; 7: 967.

Edmonds DT. Possible mechanisms for biological effects ofweak ELF electromagnetic fields, in *Bioelectrodynamics and Biocommunication,* M. W. Ho, F. A Popp, and U. Warnke, eds., World Scientific, Singapore, 109-130, 1994.

Enaki NA, Koroli VI, Bazgan S, Nistreanu A et al. Quantum Information Processes in Protein Microtubules of Brain Neurons. 3rd International Conference on Nanotechnologies and Biomedical Engineering, Volume 55 of the series IFMBE Proceedings pp 245-249, doi: 10.1007/978-981-287-736-9_60., 2016

Fantappiè L. Principi di una teoria unitaria del mondo fisico e biologico, Di Renzo Editore, Roma, 1993.

Firstenberg O., Peyronel T., Qi-Yu Liang, Gorshkov A.V., Lukin M.D. and Vuletić V., "Attractive photons in a quantum nonlinear medium", Nature 2013; 502: 71–75. doi:10.1038/nature12512.

Fiscaletti D and Sorli A. About a three-dimensional quantum vacuum as the ultimate origin of gravity, electromagnetic field, dark energy ... and quantum behaviour. Ukrainian Journal of Physics 2016; 61(5): 413-431.

Ghosh S et al. Evidence of Massive Global Synchronization and the Consciousness: Comment on

"Consciousness in the Universe: A Review of the 'Orch OR' Theory" by Hameroff and Penrose, Phys Life Rev 2013; 11(1): 83-100.

Hagelin JS, Rainforth MV and Cavanaugh KLC. et al. Social Indicators Research 1999; 47: 153. doi:10.1023/A:1006978911496.

Hu H and Wu M. Spin-mediated consciousness theory. arXiv 2002; quant-ph/0208068. Also see Med. Hypotheses, 2004; 63: 633-646.

Hu H and Wu M. Spin as primordial self-referential process driving quantum mechanics, spacetime dynamics and consciousness. NeuroQuantology, 2004; 2:41-49.

Hu H and Wu M. Concerning spin as mind-pixel: How mind interacts with the brain through electric spin effects, NeuroQuantology, 2008; 6(1): 26-31.

Kwok S and Zhang Y. Mixed aromatic–aliphatic organic nanoparticles as carriers of unidentified infrared emission features. *Nature* 2011; DOI: 10.1038/nature10542

Jenkins JM, Twicken JD, Batalha NM, Caldwell DA, Cochran, William D, Endl M, Latham DW, Esquerdo GA, Seader S, Bieryla A, Petigura E, Ciardi DR, Marcy GW, Isaacson H, Huber D, Rowe JF, Torres G, Bryson ST, Buchhave L, Ramirez I, Wolfgang A, Li J, Campbell JR, Tenenbaum P, Sanderfer D, Henze CE, Catanzarite JH, Gilliland RL, Borucki WJ. Discovery and Validation of Kepler-452b: A 1.6 R⊕ Super Earth Exoplanet in the Habitable Zone of a G2 Star. The Astronomical Journal 2015; 150(2): 56.

Lazar SW et al. Meditation experience is associated with increased cortical thickness, Neuroreport. 2005; 16(17): 1893–1897.

Lipton BH. The Biology of Belief: Unleashing the Power of Consciousness, Matter, & Miracles, Amazon, 2007.

Meijer DKF and Geesink HJH. Phonon guided biology: architecture of life and conscious perception are mediated by toroidal coupling of phonon, photon and electron information fluxes at discrete eigenfrequencies. Neuroquantology 2016; 14(4): 718-755.

Moldovanu I and Vovc V. A new conceptual dimension of functional neurology: altered states of consciousness. Bul. Acad. Sci. of Moldova. Medical Sciences 2013; 2(38): 8–15.

Pereira C. The cosmic energy bridge, cellular quantum consciousness and its Connections. J Metaphysic Connect Conscious 2015.

Pike OJ, Mackenroth F, Hill EG and Rose SJ. A photon–photon collider in vacuum hohlraum. Nature Photonics (2014); 8:434–436.

Persinger MA, Dotta B T, Saroka KS and Scott MA. Congruence of energies for cerebral photon emissions, quantitative EEG activities and ~5 nT changes in the proximal geomagnetic field support spin-based hypothesis of consciousness. Journal of Consciousness Exploration & Research, 2013; 4(1): 1-24.

Radin D. Et al. Compassionate intention as a therapeutic intervention by partners of cancer patients: effects of distant intention on the patients' autonomic nervous system. Explore 2008; 4: 235-243.

Reddy JSK and Pereira C. Origin of life: a consequence of cosmic energy, redox homeostasis and quantum phenomenon. NdeuroQuantology 2016; 14(3): 581-588.

Rosenkranz M, Davidson RJ, MacCoon D et al. A comparison of mindfulness-based stress reduction and an active control in modulation of neurogenic inflammation. *Brain, Behavior, and Immunity* 2013; 27(1):174–184.

Rubia K. The neurobiology of meditation and its clinical effectiveness in psychiatric disorders. *Biological Psychology* 2009; 82(1):1–11.

Satyajit S, Subrata G, Batu G, Krishna A, Kazuto H, Daisuke F and Bandyopadhyay A. Atomic water channel controlling remarkable properties of a single brain microtubule: Correlating single protein to its supramolecular assembly. Biosensors and Bioelectronics 2013; 47: 141–148.

Schrödinger E. What is life?, Cambridge University Press, 1944.

Sorli A. Advanced Relativity, The unification of matter, space, mind, and consciousness, Amazon, 2016.

Sorli A, Fiscaletti D and Mageshwaran M. Advanced Relativity, Unification of matter, space and consciousness. NeuroQuantology 2016; 14(4): 645-656.

Sun Y, Wang C and Dai J. Biophotons as neural communication signals demonstrated by in situ biophoton autography, Photochem Photobiol Sci. 2010; 9(3): 315-322.

Tang R and Dai J. Biophoton signal transmission and processing in the brain, J Photochem Photobiol B. 2014; 139:71-5.

Tarlaci S. Spin-mediated consciousness theory. NeuroQuantology 2006; 1: 32-44.

Tarlaci S. Why we need quantum physics for cognitive neuroscience and psychiatry. NeuroQuantology 2010; 8(1): 66-702421.

Tart C, Puthoff H and Targ R. Information Transmission in Remote Viewing Experiments. Nature 1980; 284: 191.

Tolle E. The Power of Now: A Guide to Spiritual Enlightenment, Namaste Publishing, 1997.

Villasante A et al. Binding of microtubule protein to DNA and chromatin: possibility of simultaneous linkage of microtubule to nucleic and assembly of the microtubule structure. Nucleic Acids Res. 1981; 9(4): 895–908.

Experiential Methodology in Consciousness Research

Abstract

Consciousness research in the last few decades is approaching consciousness with the classical research methodology, namely, as consciousness was an object similar to quantum fields or elementary particles. This approach is not grasping the subjective aspect of consciousness, namely, one has the ability to observe, to be conscious of how one's mind is creating a scientific model of consciousness. Experiential methodology focuses on the subject which is observing the mind working on the model of consciousness. It allows you to experience the source of the observation, which seems is consciousness itself. Consciousness research requires the enlargement of scientific research methodology in which experience has the same scientific validity as the measurement. We cannot directly measure the experience of the source of our observation, however, we can experience it and our experience is real. Conceiving a subjective experience as "non-scientific" is the main obstacle of today's science because it is excluding "the observer - the one who has the experience" out of the scientific picture of the world. The aim of this article is to include human experience as data whose value is of the same ontological importance as data obtained by measurement.

Key words: conscious observer, scientific methodology, consciousness

1. **Introduction**

Science of the 20th century has considered that the scientific picture of the world is "objective" and so something which is real. Recent epistemological research is showing that the scientific picture of the world is only "rational", one cannot claim that it is "real" (Magershwaran et al., 2016). In science we experience the world through scientific models which are rational pictures of the world produced by the scientific human mind. The ultimate reality of the world is far beyond this rational picture.

In the "scientific model" of today's science, only phenomena that can be measured are entered. Certain individual human experiences cannot be measured and remain out of the research frame of today's science. They are described as "non-scientific" only because they are outside of the grasp of today's scientific methodology. This fact is a serious cognitive and ethical problem in today's science; it will be stressed in this article and a new enlarged methodology will be proposed which includes all the phenomena that the observer can experience.

2. **Materials and methods**

The observer is the core of today's physics. It is widely accepted that the observation of a given superposition favours the superposition to manifest in the physical world. Less known, but not of less importance, is that the observation gives the duration to the observed motion in space. Without the observation, there is no duration (Fiscaletti, Sorli., 2015a). We cannot measure these crucial impacts of the observer on scientific methodology; however, we cannot deny their existence. At this point, it is important to discover the origin of the observer. The search for the observer's origin cannot be carried out by classical scientific methods; we will employ the observer's capability to observe, to "watch", "witness" different layers of reality. The observer is able to watch physical reality and is able to watch the models of physical reality which have been built by his scientific mind. He can watch the space in which there are physical objects and he can watch space in which there are mental objects. If the outer and inner space is the same space, there remains an open question which is not the core of this article. It is important that the observer can observe both of the spaces which means the observer has a higher ontological status than the inner and outer space. The observer and the source of observation are beyond the space.

The research methodology on the origin of the observer is the following:

1. You watch and become conscious of the physical environment in which you are at the moment
2. You watch and become conscious of the space in which there are physical objects
3. You close your eyes, watch and become conscious of mental objects
4. You watch and become conscious of the space in which the mental objects are.

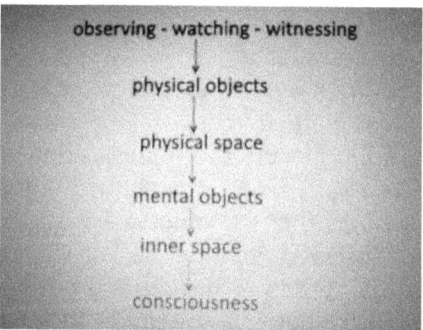

Figure 1: Experiential methodology in consciousness research

Experiential methodology is based exclusively on your perception, there is no theory behind it. You cannot measure the act of observation, however, this does not mean that it is "non-scientific". The observation is the core of science and this has to be fully acknowledged. By practising this methodology regularly you will discover that consciousness is beyond the inner space in which the mental phenomena exist. In the discussion section, we will place the experience of consciousness into the model of Advanced Relativity, being aware that the model is not exactly describing the experience, it is only pointing to it.

3. Discussion

With the acceptance that the experience of the observer is of the same importance as the measurement scientific methodology and is enlarged and able to embrace the entire existence. The observer who is conscious that his scientific observation happens through his scientific mind's picture of the world has a huge potential to develop a scientific picture of the world which will be more adequate than the world itself.

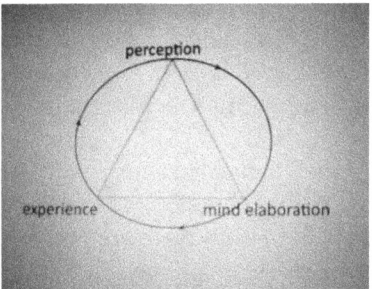

Figure 2: The epistemology triangle of the conscious observer

The conscious observer is aware that the model of linear time "past-present-future" is only a psychological time through which the common observer experiences a stream of changes which run in space. The conscious observer is aware that the "space-time" model has no counterpart in physical reality. The hundred year-old conviction that time is the 4th dimension of space has ended (Fiscaletti, Sorli., 2015b).

In space it is always NOW; the time we measure with clocks is merely a numerical order of material changes, i.e. motion in space. This view of time is the basis for the renaissance of the cosmology and evolution of life which are running in the NOW, and is Einstein's fulfilled vision of time expressed in the famous quote: "...there is something essential about the NOW which is just outside the realm of science. People like us, who believe in physics, know that the distinction between the past, present, and future is only a stubbornly persistent illusion. Time has no independent existence apart from the order of the events by which we measure it." The old view, namely, that the universe and life evolve in some physical time is wrong and should be replaced with the view which fully corresponds to the scientific perception: events run in space only and time when their duration is measured (Sorli., 2017a).

Today's scientific models of consciousness are placed in the theoretical frame of space-time, where time is considered to be a physical quantity. The Hameroff-Penrose model, for example, is considering that consciousness consists of discrete events: "Thus, we may argue that consciousness consists of discrete events at varying frequencies occurring across regions of the brain, for example 40 conscious moments per second, synchronized among the neurons in the frontal and parietal cortex. What are these conscious moments?" (Hameroff, Penrose., 2014).

In the Advanced Relativity (AR) model, consciousness is not an event which has any duration. Consciousness is a fundamental n-dimensional Hilbert space in which the mind exists (described with lower dimensional Hilbert spaces) and finally, 3D physical objects do exist (Sorli at al., 2017b). When the observer is using a clock and measuring the numerical order of events running in the consciousness appears as the duration of events. This inside, in the real nature of time, is the first rational step towards experiential consciousness research. The AR model has a clear answer to the Hameroff-Penrose question: "What are these conscious moments?" These conscious moments are the experience of the observer which is still locked in the concept of linear time which is his own mind's creation. For the conscious observer, no event has its inherent duration because it is running in the consciousness where it is always and only NOW. The duration is the result of the observer's measurement.

The experiential methodology is a valuable solution for Chalmers "Hard problem of consciousness": "The really hard problem of consciousness is the problem of experience. When we think and perceive, there is a whirl of information-processing, but there is also a subjective aspect" (Chalmers., 1995). The subjective aspect of the experience is resolved by knowing who is having the experience. We cannot reach into the core of the experience by using a classical scientific tool which is measurement. By adding experience as a valid scientific tool, the hard problem of consciousness is resolved. In "brain-mind-consciousness" research we can continue measuring and mapping the brain with an encephalogram and searching for the relationship between the encephalogram maps and higher cognitive processes. Including experiential methodology we will also be able to see clearly how related the brain, mind and consciousness are:

$$measuring + experience = science\ of\ consciousness.$$

In consciousness research we need to take in account that scientific apparatus are and will probably remain 3D and will never be able to measure mind and consciousness which are multidimensional. Rapid development of computers in last decade is promising development of "artificial intelligence" which is a wrong term. 3D apparatus can only develop "logical thinking" which is far below "intelligence". Intelligence is the result of cognitive activity of the mind in higher dimensional Hilbert spaces. One could say that intelligence means logic which is inspired with consciousness:

$$\text{logic} + \text{consciousness} = \text{intelligence}$$

Consciousness is and will remain the domain of living organisms. Without fully recognizing this fact, the entropy of human society (violence, poverty, tensions between different nations, religions and cultures) will inevitably increase. Technology is not the solution for today's world. The solution is in the introduction of an experiential methodology on the origin of the observer in the worldwide educational system. On the surface we have different social backgrounds, different cultures and religions, but deep inside we are all one consciousness. By deepening in consciousness, peace grows within. The peace between humans has its basis in the inner peace of each individual. We cannot "fight" for peace; we can only merge into peace which is consciousness.

The observer's deepening into consciousness leads into the non-dual experience. Consciousness is beyond the duality "observer" and "observed". In Advanced Relativity, consciousness is the unified field in which the entire physical and mental universe exists. The conscious observer which has fully merged into consciousness experiences mental and physical objects as one with him. The big gap between "me" and the "world" which is characteristic for a rational scientific experience turns into a deep relationship of oneness. This oneness is known to all who have attained a non-dual experience. In Buddhism this is called "Shunyata", in Hinduism "Brahma", in Taoism "Tao". Each culture has its own tradition of merging into consciousness. Science does not fully understand non-dual experience yet because of the narrow methodology based exclusively on measurement. Advanced Relativity has a model of reality where the non-dual experience is included and is the most valuable and noble scientific achievement. Albert Einstein used to call it "mysterious". His famous quote is like this: "The most beautiful thing we can experience is the mysterious. It is the source of all true art and science. He to whom the emotion is a stranger, who can no longer pause to wonder and stand wrapped in awe, is as good as dead - his eyes are closed. The insight into the mystery of life, coupled though it be with fear, has also given rise to religion. To know what is impenetrable to us really exists, manifesting itself as the highest wisdom and the most radiant beauty, which our dull faculties can only comprehend in their most primitive formsthis knowledge, this feeling is at the centre of true religiousness." (Howard W. Eves.,1977).

From the ontological point of view, non-dual experience is possible, because the mental and physical world are the structures of consciousness. In most spiritual traditions, consciousness is the "stuff" out of which the mental and physical world are made. Every mental or physical form exists in the consciousness and is made out of it. This is also the view of the "Unified Reality Theory" model according to which the universe is developing itself in order to have a self-experience of its source which is consciousness (Kaufman., 2002). Advanced Relativity (AR) is familiar with this view: consciousness is the photon with an infinite frequency and exists in n-dimensional Hilbert space. By lowering the frequency in lower dimensional Hilbert spaces, this "photon-consciousness" becomes mind and finally in 3D reality it becomes matter (Sorli et al., 2016). The matter has an inherent tendency to develop in life and further in intelligent and organisms in the entire universe because matter exists in the consciousness (Sorli at al., 2017c).

It makes sense to search for an adequate scientific model of consciousness and at the same time it is important to know that the model is limited because consciousness is, from the ontological perspective, higher than the mind. The mind can only build a model of consciousness to show the path, the rest is individual research in which experiential methodology is an indispensable tool.

4. Conclusions

Today's science of consciousness needs a new paradigm which will be built primarily on the human perception and experience, and secondary on the scientific models of consciousness. The very nature of consciousness is subjective and that's why subjective experience is the most relevant result. With the introduction of experiential methodology as a valid scientific tool, science will enlarge the area of research which is essential for the development of science and which will search for the good of humanity and the entire life on the planet.

References:
Chalmers, D., Facing Up to the Problem of Consciousness, *Journal of Consciousness Studies* 2(3):200-19, 1995
Fiscaletti, D. and Sorli, A.: "Perspectives of the Numerical Order of Material Changes in Timeless Approaches in Physics", Foundations of Physics 45, 2, 105-133 (2015a).
Fiscaletti D and Sorli A. Bijective epistemology and space-time. Foundations of Science 2015b; 20(4): 387-398.
Kaufman, S., Unified reality theory: the evolution of existence into experience, Destiny Toad Press, 2002
Magi Mageshwaran, Amrit Sorli, Davide Fiscaletti. The Foundations of the Epistemology and the Methodology of Physics. *American Journal of Modern Physics*. Special Issue: Insufficiency of the Big Bang Cosmology. Vol. 5,

No. 4-1, 2016, pp. 27-33. doi: 10.11648/j.ajmp.s.2016050401.15

Hameroff S., Penrose R., Consciousness in the universe. A review of the 'Orch OR' theory, Physics of Life Reviews 11 (2014) 39–78

Howard W. Eves, In Mathematical Circles, Boston: Prindle, Weber & Schmidt, (1969)

Sorli A., Fiscaletti D. Mageshwaran M., Advanced Relativity: Unification of Space, Matter and Consciousness, NeuroQuantology, Vol 14, No 4 (2016)

Sorli A., Advanced Relativity: Bijective Physics, Amazon (2017a)

Sorli A, Kaufman S., Dobnikar U. Fiscaletti D., Advanced Relativity for the Renaissance of Cosmology and Evolution of Life (in press NeuroQuantology 2017b)

Sorli A. Dobnikar U., Fiscaletti D., Koroli V., Advanced Relativity: Multidimensionality of Consciousness and Mind, Origin of Life, Psi Phenomena, NeuroQuantology, Vol 15, No 2 (2017c).

Unified Field Theory Based on Bijective Methodology

Abstract

Einstein's dream was to describe all fundamental forces by an unique field. This article is an attempt to realise this Unified Field Theory (UFT) using the bijective methodology, which confirms that both the particles and fields cannot exist in an empty space, deprived of physical properties. Space is a 4D continuum as denoted by Einstein; space is the fundamental energy of the universe, a superfluid in which energy is syntopic. Electric field is excitation of this 4D superfluid by coordinates X1,X2,X4. Magnetic field is excitation of this 4D super-fluid by coordinates X1,X3,X4. Bio-field (also called "morphogenetic field", "QI" in Eastern medicine) is excitation of this 4D super-fluid by coordinates X1,X2,X3,X4. Strong nuclear force and gravity have origin in the diminished energy density of this 4D superfluid caused by the presence of massive particles and massive objects.

Key Words: unified field theory, bijective methodology, quantum vacuum, super-fluid space.

Introduction of bijective methodology

The concept of "Bijective methodology" introduces a new research methodology in physics, which follows bijective analysis of physical equations, where each element of a given model corresponds to the exactly one element in physical reality. Applying bijective methodology we will get confirmation that both the particles and fields exist in space. For more than 100 years, in physics we are convinced that space is empty and deprived of all physical properties. As particles and fields have physical properties; so the space in which they exist must have some physical properties. The "empty space" is an element in physical models which has no existence in physical reality and should be abandoned from physics (*Sorli, 2018a*).

We have introduced in this article space as a "4D super-fluid", with the energy density of which has the value of Planck energy density: $4.633 \cdot 10^{113} J/m^4$. As the space is considered as 4D, so we will use m^4 instead of m^3. Einstein has defined space as 4D continuum, but somehow his vision was not fully understood till these days. Despite Einstein model of 4D continuum it is still not usual to imagine that the space has four dimensions and that space as the 4D super-fluid with its characteristic energy density.

The bijective methodology, which we have proposed (*Sorli, Patro, 2018*) allows the creative an adequate imagination which is based on human perception and experimental data. This methodology excludes the possibility of an error in modelling models of reality: it gives giving more credibility to the creative imagination based on direct or instrumental perception, rather than to use pure mathematical speculations.

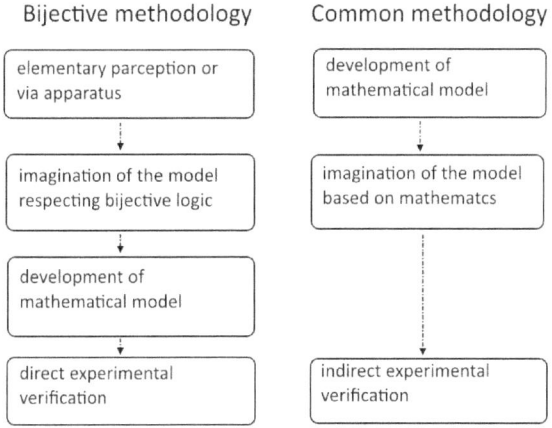

Figure-1 Bijective methodology and common methodology in today physics

By using bijective methodology, we have discovered that imagine of "empty space" in which particles and fields exist is the wrong imagination. Particles and fields are manifestation of the 4D superfluid space which has concrete physical properties, namely, its energy density.

Common methodology of today physics is firstly development of mathematical model which is not based on perception and further development of wrong imagination which is proved with an indirect experimental verification. Good example of common research methodology of today physics is development of the Higgs mechanism which is based on super-symmetry model SUSY which predict all particles are mass-less. SUSY is against of "mass-energy equivalence principle" which is experimentally directly proved and is one of pillars of physics. On the base mechanism of SUSY wrong imagination is born, namely, that must exist a field which gives mass to the particles. This wrong imagination is finally confirmed with the discovery of Higgs boson which does not prove existence of Higgs field. Higgs boson is artificial man made particle which has no existence in the universe on its own and cannot be seen as a proof for the Higgs mechanism (Sorli, 2018a).

Other classical example of common methodology based on wrong imagination is the idea of holographic mass which also has no correspondence in physical reality. Haramein introduces in his model a new physical entity called "Planck spherical unit": In order to better represent the natural systems of harmonic oscillators we initiate our calculation by defining a Planck spherical unit (PSU) oscillator of the Planck mass m_l with a spherical volume V_{ls} and a Planck length diameter $l = 1{,}616199 \cdot 10^{-33} cm$ with a radius of $l_r = \dfrac{l}{2}$. We utilize a spherical volume for our fundamental spacetime quantum foam PSU oscillator instead of the typical Planck area l^2 or Planck volume l^3 in our generalized holographic approach. Therefore a spherical PSU of radius l_r has a volume of $2{,}210462 \cdot 10^{-99} cm^3$ (Haramein, 2013). In continuation of his article we can read: Therefore, we find that the number of discrete Planck masses within any given mass m multiplied by $2l$, which is a discrete quantity, will generate the holographic radius equivalent to the well known Schwarzschild radius of equation (1) so that in the case of equation (19) we have a non-relativistic form derived from discrete vacuum oscillator Planck quantities generating a quantized solution (Haramein, 2013).

According to Haramein statement above, one can write following equation:

$$n \geq 1 \rightarrow m = n \cdot m_p \cdot 2l \qquad (1),$$

where m is any given mass, l is a Planck distance, n is the discrete natural number of Planck masses in a given mass m. Number n can be 1, 2 or higher natural numbers.

Rearranging equation (1) we get:

$$n = \frac{m}{m_p \cdot 2l} \quad (2).$$

Equation (2) is has no physical meaning because on the left side we have unit in kg and on the right side we have unit kgm. When we put values of proton mass which is $1,6726219 \times 10^{-27} kg$ we get following value for n:

$$1,6726219 \cdot 10^{-27} kg = n \cdot 2,17647051 \times 10^{-8} kg \cdot 2 \cdot 1,616199 \cdot 10^{-35} m$$

Or, $$n = \frac{1,6726219 \cdot 10^{-27} kg}{(2,17647051 \cdot 10^{-8} kg)(3,232398 \cdot 10^{-35} m)}$$

$$n = 2,377494 \cdot 10^{15},$$

which is away from discrete natural numbers for the magnitude error of 10^{15}. How from equation (1) one arrives to the "holographic radius" equivalent to the well known Schwarzschild radius remains unknown.

Other Haramein imagination which is not passing bijective methodology and calculations is that the energy of space inside the proton volume is equal to the sum of masses of all protons in the observable universe. Proton volume: $V_P = 2,5 \cdot 10^{-45} m^3$. Planck energy density of space: $\rho_{PE} = 4,633 \cdot 10^{113} J/m^3$

$$E_{of\,.space.in.\,proton\,.volume} = 4,633 \cdot 10^{113} J/m^3 \cdot 2,5 \cdot 10^{-45} m^3$$

$$E_{of\,.space.in.\,proton\,.volume} = 1,158 \cdot 10^{69} J/m^3$$

Mass of the stellar objects which we can consider mass of all protons in observed universe is $10^{53} kg$. Energy of stellar objects of observed universe is $10^{53} kg \cdot c^2$ which yields:

$$E_{stars.of.the.universe} = 10^{53} kg \cdot 8,99 \cdot 10^{16} m^2 s^{-2}$$

$$E_{stars.of.the.universe} = 10^{53} kg \cdot 8,99 \cdot 10^{16} m^2 s^{-2}$$

$$E_{stars.of.the.universe} = 8,99 \cdot 10^{69} J$$

The difference between the amount of protons of the universe calculated as energy and the amount of space energy in the volume of the proton is following: $8,99 \cdot 10^{69} J - 1,158 \cdot 10^{69} J = 7,832 \cdot 10^{69} J$, which is error of magnitude 10^{69}.

In Haramein model space is imagined as the structure made out of Planck spherical units (PSU) which are expressed in unit of kg (kilogram). We measure in physics with kilogram amount of matter of a given physical object. Space is not matter; space is pure energy which one can adequately imagine as the 4 dimensional superfluid which has its characteristic energy density, namely, Planck energy density.

Higgs mechanism and holographic mass are classical examples of models which are based exclusively on the abstract speculations and have no support in direct observation and measurement. Bijective methodology has the ability to end the era of this "speculative physics" which has lost connection with the real world.

In line with Einstein's view of completeness of a physical theory, according to bijective methodology, one requires that a bijective correspondence between elements of the model and elements of physical reality must exist, where, at a fundamental level, elements of physical reality can be defined as those elements which have a primary physical existence, namely either are perceived directly by senses (without the necessity to make measurements) or are perceived indirectly by "enhanced senses" where the adjective "enhanced" is meant to harbour a perception through, for example, radio telescopes, Geiger-Muller elementary-particle counters or electromagnetic waves detectors. In Bijective Epistemology there is a direct epistemological correlation between each element of the physical model and one element of phenomena (Fiscaletti and Sorli, 2015).

Development of Unified Field Theory

In Einstein's Unified Field Theory (UFT), the idea is to unify all the four fundamental forces: electromagnetic, strong and weak nuclear and gravitational force. The term 'force' that we have in physics since Newton:

$$F = m \cdot a \quad (3).$$

In equation (3) above, force F denote the force with which material object with the mass m which has acceleration a hits the wall. After the hit, the force F is gone, acceleration a is gone, only mass m remains. The common imagination in physics today is that force F exists on its own as for example the mass m. Bijective model confirms that this type of imagination is absolutely wrong. The force F of the moving object with mass m exists as a physical property of the moving object. It does not have existence on its own.

Introduction of 'forces' in quantum physics is quite not right, because we think in terms of such an identity, which do not exist. There are no forces as 'electromagnetic force', 'gravitational force', 'strong nuclear force' and 'weak nuclear force'. We can only search on the following phenomena: 1. electromagnetism, 2. macro gravity between physical objects, micro gravity between protons and neutrons inside the nucleon of the atom and quantum gravity regarding the gravitational interaction at the Planck scale (where general relativity has to be embedded with the laws of quantum theory), 3. radiation by nucleus decay (also called "radioactive decay"). Bijective analysis of Special Relativity Theory confirms (Sorli, 2018a) that universal space is not 3D + T (three spatial and one temporal dimension), universal space is the **4 d**imensional **s**yntropy **s**pace where time is mathematical parameter of changes running in space (which we will call from now on **4DSS**). 4DSS can be considered as the real origin and source of electromagnetism, macro gravity, micro gravity and quantum gravity as well as of radioactive decay, which indeed can be seen as different aspects of the same coin. Moreover, introducing 4DSS offers original explanation of light constancy, namely, the source of the light is moving in 4DSS and all inertial systems are moving in 4DSS. That's why light has the same speed in all inertial systems regardless of the motion of its source and obeys Doppler law.

It will be shown in this article that 4DSS is the only field which exists in the universe. Macro gravity, micro gravity inside the nucleus between protons and neutrons, quantum gravity regarding the gravitational interaction at the Planck scale, radioactive decay, electromagnetism and also bio-field will be described as different phenomena of the same fundamental background, which is the 4DSS.

Macro gravity and micro gravity inside the nucleus

Gravity has origin in diminished energy density of 4DSS which corresponds to amount of the energy of a given physical object according to the following formula:

$$E = mc^2 = (p_{PE} - \rho_{SE}) \cdot V \quad (4),$$

where E is the energy of the given physical object, m is its mass, V is its volume, ρ_{PE} is Planck energy density and ρ_{SE} is the diminished energy density in the centre of the object. Gravity has origin in the area of diminished energy density of 4DSS in which two physical objects are "captured". Outer higher pressure of 4DSS is pushing in the direction towards the areas of lower energy density of 4DSS.

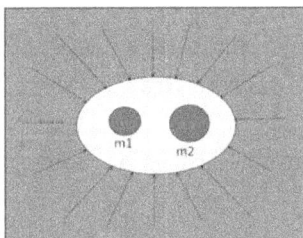

Figure 2 (Gravity as the result of outer pressure of 4DSS)

There is no 'carrier' of gravity between two physical objects. Objects are 'captured' in the area of lower energy density of 4DSS and pushed to each other via difference of outer and inner pressure of 4DSS.

In the centre of proton vortex energy density of 4DSS is smaller regarding Planck energy density for $6,0 \cdot 10^{34} J/m^4$. In the centre of the planet Earth energy density of 4DSS is smaller regarding Planck energy density for $4,9 \cdot 10^{20} J/m^4$ (Sorli, 2018a). Proton is diminishing energy density of 4DSS for scale 10^{14} more than our planet Earth. As proton diameter is extremely small as compared to the Earth diameter, the diminished area of 4DSS is also small and limited inside the atom nucleus where protons and neutrons are pushed together because of the higher outer pressure of 4DSS.

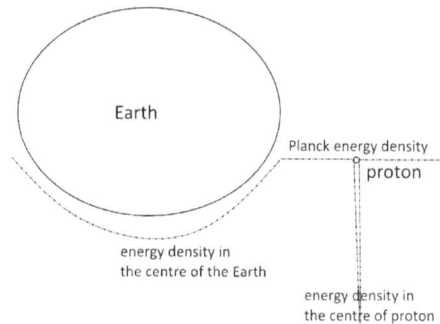

Figure 3 (Diminished energy density of 4DSS of the Earth and of the proton)

Diameter of atom nucleus is about 6 femto-meter where one femto-meter is $10^{-15} m$. At the range of $10^{-15} m$ the 'strong force' between protons inside the nucleus is about 10^{38} times as strong as gravitation. This 'strong force' is micro gravity which is characteristic only inside the nucleus.

4DSS is physical quantum vacuum

According to quantum mechanics of 20th century, the quantum vacuum state is not truly empty but instead contains fleeting electromagnetic waves and particles that pop into and out of existence. In our view quantum vacuum excitation are electromagnetic waves and particles are vortexes of quantum vacuum which we give new name 4DSS (4 dimensional syntropy space), which exactly express its physical properties.

4DSS defined by a fundamental variable energy density associated to elementary (reduction-state) RS processes of creation/annihilation of quanta, giving rise to a total zero spin, thus constituting an organized superfluid Bose ensemble, is the real origin of all subatomic particles, such as electrons, positrons, photons, hadrons etc. as well as of all macroscopic bodies (Fiscaletti and Sorli, 2016 and 2018). The appearance of baryonic matter derives from an opportune excited state of the 4DSS defined by opportune fluctuations of its energy density and corresponding to specific reduction-state (RS) processes of creation/annihilation of quanta of this superfluid vacuum. The excited state of the superfluid quantum vacuum corresponding to the appearance of a material particle of mass m obtained from equation (4) is determined by opportune RS processes of creation/annihilation of quanta described by a wave function $C = \begin{pmatrix} \psi \\ \phi \end{pmatrix}$ at two components satisfying a time-symmetric extension of the Klein-Gordon quantum relativistic equation

$$\begin{pmatrix} H & 0 \\ 0 & -H \end{pmatrix} C = 0 \qquad (5)$$

where $H = \left(-\hbar^2 \partial^\mu \partial_\mu + \frac{V^2}{c^2}(\Delta\rho_{qvE})^2\right)$ and $\Delta\rho_{qvE} = (\rho_{PE} - \rho_{qvE})$ is the change of the quantum vacuum energy density. Equation (17) corresponds to the following two equations:

$$\left(-\hbar^2 \partial^\mu \partial_\mu + \frac{V^2}{c^2}(\Delta\rho_{qvE})^2\right)\psi_{Q,i}(x) = 0 \quad (6)$$

for creation events and

$$\left(\hbar^2 \partial^\mu \partial_\mu - \frac{V^2}{c^2}(\Delta\rho_{qvE})^2\right)\phi_{Q,i}(x) = 0 \quad (7)$$

for destruction events. As a consequence of the motion of the virtual particles corresponding to the elementary fluctuations of the energy density the superfluid 4F space is filled with virtual radiation with frequency

$$\omega = \frac{2\Delta\rho_{qvE} V}{\hbar n} \quad (8).$$

This means that each elementary fluctuation of the quantum vacuum energy density in a given volume corresponds to a vibratory oscillation of the vacuum at a peculiar frequency, in particular that each material object can be interpreted as the result of specific vibratory states of the vacuum.

Moreover, by decomposing the real and imaginary parts of the Klein-Gordon equation (5) after writing the two components of the wave function in polar form

$$\psi_{Q,i} = |\psi_{Q,i}| \exp\left(\frac{i S^\psi_{Q,i}}{\hbar}\right) \quad (9),$$

$$\phi_{Q,i} = |\phi_{Q,i}| \exp\left(\frac{i S^\phi_{Q,i}}{\hbar}\right) \quad (10),$$

for the real part one obtains a couple of quantum Hamilton-Jacobi equations that, by imposing the requirement that they are Poincarè invariant and have the correct non-relativistic limit, assume the following form

$$\partial_\mu \begin{pmatrix} S^\psi_{Q,i} \\ S^\phi_{Q,i} \end{pmatrix} \partial^\mu \begin{pmatrix} S^\psi_{Q,i} \\ S^\phi_{Q,i} \end{pmatrix} = \frac{V^2}{c^2}(\Delta\rho_{qvE})^2 \exp\begin{pmatrix} Q^\psi_{Q,i} \\ -Q^\phi_{Q,i} \end{pmatrix}$$

(11),

while the imaginary part gives the continuity equation

$$\partial_\mu \left(\sigma \partial^\mu \begin{pmatrix} S^\psi_{Q,i} \\ S^\phi_{Q,i} \end{pmatrix} \right) = 0 \quad (12)$$

where σ is the ensemble of particles associated with the wave function under consideration and

$$Q_{Q,i} = \frac{\hbar^2 c^2}{V^2(\Delta\rho_{qvE})^2} \left(\frac{\left(\nabla^2 - \frac{1}{c^2}\frac{\partial^2}{\partial t^2}\right)|\psi_{Q,i}|}{|\psi_{Q,i}|} - \frac{\left(\nabla^2 - \frac{1}{c^2}\frac{\partial^2}{\partial t^2}\right)|\phi_{Q,i}|}{|\phi_{Q,i}|} \right)$$

(13)

is the quantum potential of the vacuum. The quantum potential of the vacuum (13), which can be considered as the ultimate entity which guides the occurring of the processes of creation or annihilation events, implies that 4DSS in the quantum regime acts in the form of an undivided non-local network of RS processes where time has not a primary physical reality but exists merely as a mathematical parameter measuring the dynamics of the particle into consideration (Fiscaletti and Sorli, 2018).

Now, from equation (11), by changing the ordinary differentiating ∂_μ with the covariant derivative ∇_μ and by changing the Lorentz metric with the curved metric $g_{\mu\nu}$, the following equations of motion for the fluctuations of the quantum vacuum energy density (which give origin to a creation event for a quantum particle Q of a given mass) in a curved background can be obtained:

$$\tilde{g}_{\mu\nu}\tilde{\nabla}_\mu S_{Q,i}\tilde{\nabla}_\nu S_{Q,i} = \frac{V^2(\Delta\rho_{qvE})^2}{c^2\hbar^2} \quad (14),$$

where $\tilde{\nabla}_\mu$ represents the covariant differentiation with respect to the metric

$$\tilde{g}_{\mu\nu} = g_{\mu\nu}/\exp Q_{Q,i} \quad (15)$$

which is a conformal metric, where

$$Q_{Q,i} = \frac{\hbar^2 c^2}{V^2(\Delta\rho_{qvE})^2} \frac{\left(\nabla^2 - \frac{1}{c^2}\frac{\partial^2}{\partial t^2}\right)|\psi_{Q,i}|_g}{|\psi_{Q,i}|}$$

(16)

is the quantum potential of the vacuum.

Equations (14) and (16) can also be expressed in terms of the vibratory states of the vacuum. In fact, by substituting the expression of the frequencies of vibration of the virtual particles (8) into equations (14) and (16), these two equations respectively read

$$\tilde{g}_{\mu\nu}\tilde{\nabla}_\mu S_{Q,i}\tilde{\nabla}_\nu S_{Q,i} = \frac{n^2\omega^2}{4c^2}$$

(17)

and

$$Q_{Q,i} = \frac{n^2\omega^2}{4c^2} \frac{\left(\nabla^2 - \frac{1}{c^2}\frac{\partial^2}{\partial t^2}\right)_g |\psi_{Q,i}|}{|\psi_{Q,i}|} \qquad (18).$$

On the basis of equations (14)-(18), one can say that the quantum potential of the vacuum determined by RS processes associated with creation events of quantum particles from the 3D quantum vacuum is equivalent to a curved 4DSS with its metric being given by (15) (or (17)). As a consequence of the specific vibratory states of the virtual particles of the superfluid medium, the quantum potential of the superfluid space giving rise to the undivided non-local network of RS processes of quanta is tightly linked with the curvature of the 4DSS. In other words, we can say that RS processes, by means of the quantum potential of the vacuum given by (16) (or the equivalent relation (18)), generate, in our macroscopic level of reality, of a curvature of the 4DSS and, at the same time, the metric of the 4DSS is linked with the quantum potential of the vacuum which rules the behaviour of the particles. In this way, a fundamental unification of quantum and gravity is obtained in a geometrical picture where the 4DSS acts as a variable energy density corresponding to elementary RS processes of a 3D timeless non-local quantum vacuum acting as a superfluid medium and endowed with virtual particles characterized by opportune vibratory states.

A crucial result of the unified field theory of the 4DSS regards the fact that the quantum potential of the vacuum ((16) or (18)) equipped with the conformal metric ((15) or (17)), by taking account fruitful considerations made by Ali and Das (2015), allows us to throw new light in the interpretation of the cosmological constant, and in particular of the grand unified theory (GUT) scale. If in Ali's and Das' approach, by assuming that the universe is filled by a condensate described by a wave function $\Psi = R\, e^{iS/\hbar}$, in the context of a Raychaudhuri equation a quantum potential of the form

$$Q = \frac{\hbar^2}{3m^2} q^{ab} \frac{\left(\nabla^2 - \frac{1}{c^2}\frac{\partial^2}{\partial t^2}\right)R}{R}$$

(19)

where $q_{ab} = g_{ab} - u_a u_b$, $u_a = \frac{\hbar}{m}\partial_a S$, leads to obtain directly a cosmological constant that, by assuming a Gaussian form of the wave function or a scalar field theory, is expressed by relation

$$\Lambda_Q = \frac{1}{L_0^2} = \left(mc/\hbar\right)^2 \qquad (20)$$

where $L_0 = \hbar/mc$ may be identified with the current linear dimension of our observable universe, m can be interpreted as the small mass of gravitons (or axions), here in our model of a unified field theory in 4DSS, in analogy to Ali's and Das' approach, by assuming a Gaussian form of the wave function of creation events $\Psi \approx \exp(-r^2/L_0^2)$, where $L_0 = \frac{\hbar c}{\Delta\rho_{qvE} V}$ is an appropriate length determined by the fluctuations of the quantum vacuum energy density, we obtain:

$$\frac{\hbar^2 c^2}{V^2(\Delta\rho_{qvE})^2}\frac{\left(\nabla^2 - \frac{1}{c^2}\frac{\partial^2}{\partial t^2}\right)|\psi_{Q,i}|}{|\psi_{Q,i}|}\bigg|_g = \frac{1}{L_0^2} = \left(\frac{\Delta\rho_{qvE} V}{\hbar c}\right)^2$$

(21).

Therefore, considering here some grand unified theory (GUT) scale φ between $10^6 J$ and $10^9 J$ one finds that the corresponding value of the quantum vacuum energy density fluctuations in the minimum quantized space given by the Planck volume is

$$\Delta\rho_{qvE} = \frac{\varphi}{l_p^3} \approx 10^{111} J/m^3 \cdots 10^{114} J/m^3 \quad (22).$$

Integration of electromagnetic quantum vacuum of QED and 4DSS

In QED photon is the excitation of electromagnetic quantum vacuum. QED is one of the most successful scientific models with exact 'bijective correspondence' with the physical world. The 4DSS has all properties of electromagnetic quantum vacuum. In this way, we unify electromagnetism and gravity, where electromagnetism is just an excitation of 4DSS and gravity is carried by 'variable energy density of 4DSS'.

The idea of 'Special theory of Relativity' that photon can move through the 'empty space' is a failure imagination which has lead to another failure imagination, namely, that gravity could be carried by a particle similar to the photon named as 'graviton'. The Photon is excitation of 4DSS and has constant speed in the area of 4DSS where energy density is the same. In stronger gravity, the 'energy density of 4DSS' diminishes minimally, which reduced the speed of photon for smaller values. Shapiro's experiment proves this model (Sorli, 2018b). We know that if we diminish the density of the medium, the velocity of the signal diminishes too.

The 4DSS is made out of the virtual photons. When these photons get in excitation by virtue of the elementary RS processes of creation/annihilation, they appear as actual photons that we observe. For example, when we hit a piece of iron, the virtual photons of 4DSS inside the iron get in excitation and radiate. We think that photons radiate from the iron which is false imagination. Photons radiate from 4DSS which is a kind of Einstein-Bose condensate, a superfluid 3D quantum vacuum. By radiating photons, the same amount of photons will remain in 4DSS. The 4DSS, through the action of the subtending 3D timeless superfluid quantum vacuum, is an infinite source of energy. The 4DSS obeys the first law of thermodynamics according to which energy cannot be created and could not be destroyed. The syntropic energy of the 4DSS represents the 'dark energy' + 'dark matter' of the universe, which is about 95% of the energy in a given volume of the universe and 5% remains of ordinary matter which is entropic energy.

The increase of the entropy of matter in the universe is only a partial process which does not increase the entropy of the universe in total; because in black holes, matter is disintegrating back into the primordial energy of the 4DSS which is made out of virtual photons (Sorli et al., 2018c). Various research works published in 'Nature Journal' confirms the idea that matter is made out of photons (Ofler at al., 2013).

Radioactive decay

In Standard Model particles and fields exist in an empty space. This imagination is the main misunderstanding of Standard Model which further leads to wrong conclusions. One of them is the interpretation of 'weak nuclear force' as a primary physical reality which is mediated by W bosons and Z bosons. We have seen that forces are epiphenomena of primordial physical processes. By radioactive decay happens that atom nucleus becomes unstable because the balance between micro gravity and repulsion is broken. Repulsion between protons and repulsion between neutrons have origin in their spinning around each other. We call these phenomena "proton-proton" and "neutron-neutron" spin pairs.

For example Standard Model understands that beta decay is a consequence of the weak force, which is characterized by relatively lengthy decay times.

Nucleons are composed of up quarks and down quarks, and the weak force allows a quark to change type by the exchange of a W boson and the creation of an electron/antineutrino or positron/neutrino pair. For example, a neutron, composed of two down quarks and an up quark, decays to a proton composed of a down quark and two up quarks. In our view W boson intended as a primary independent physical reality is not causing beta decay; it is only the result of this decay which has the cause in the instability of the nucleons. It is the specific values of the energy density of the superfluid 4D space which to our eyes seem to lead to the appearance of a W boson (or, equivalently, of a Z boson).

From this perspective W bosons and Z bosons whose lifetime is about $3 \cdot 10^{-25} s$ are not carriers of "weak nuclear force". Rather, W bosons and Z bosons are momentary fluxes of energy of 4DSS characteristic of the radioactive decay.

Integration of Biofield in physics

The term *biofield* was proposed in 1992 by an ad hoc committee of CAM practitioners and researchers convened by the newly established Office of Alternative Medicine (OAM) at the US National Institutes of Health (NIH) (Beverly at al. 2015). The 'Biofield' is the word chosen to describe the field of energy and information that surrounds and interpenetrates the human body. It is composed of both measurable electromagnetic energy and hypothetical subtle energy. This structure is also called the Human Energy Field or Aura which can be photographed by GDV camera (based on Kirlian effect) developed by Prof. Konstantin Korotkov (*Korotkov, 2018*). In Chinese traditional medicine, the bio-field is called Qi, in Indian traditional medicine (ayurveda), the bio-field corresponds to

"Prana". Science of 20th century has ignored the existence of 'QI' or 'Prana', which has millennium of tradition. Science of 21th century is requiring to incorporate the 'biofield', in order to develop more adequate picture of reality.

Phenomenon of 'Biofield' is still missing coherent scientific interpretation which we will present in this chapter. The 'Biofield' is composed out of 4D photons which are excitation of 4DSS. The 'Biofield' has not only 4D structure but also higher dimensional structures which we describe with higher dimensional Hilbert spaces. The 'Biofield' functioning is governed via higher dimensional structures by consciousness itself (*Sorli et al., 2017*).

Preliminary experiments confirm that 'biofield' as a subtle energy structure additionally decreases energy density of space which causes that 'living mass' has more weight than the same 'dead mass' (*Sorli, 2002*). Preliminary experiments has been carried out at the Bio-technical faculty, Ljubljana, Slovenia in June 1987. Measurements have been performed on a Mettler Zurich M5 scale. Six test-tubes were filled with three milli-liters of a water solution made out of meat and sugar. Four test-tubes were used and a fungus was put into two of the test-tubes. All of test tubes were welded airtight. The weight difference between test-tubes was measured for ten days. After three days of growth, the weight of test-tubes with the fungus increased (on average) 34 micrograms and in last seven days remains unchanged. The experiment was carried out in sterile circumstances. Here the biomass is increasing by incorporating nonliving substances and could be represented by the following equation:

$$m_{organic} + \Delta m = m_{living} \quad (23),$$

where $m_{organic}$ is the mass of the organic matter, Δm is the change in mass of the system, and m_{living} is the mass of the living organisms which have transformed organic matter in living matter.

In another experiment, two test-tubes were filled with 5 grams of Californian worms with distilled water. All of the test-tubes were then welded airtight. The weight difference between test-tubes was measured for 5 hours. At the end of the first hour there was no appreciable difference but at the end of the second and third hour there was an increased mass of 4.5 micrograms on average. This mass then remained stable for the next 2 hours most likely due to there no longer being any living organisms. This change in mass due to the change of organisms from a living condition to a nonliving one could be shown with the following equation:

$$m_{living} = m_{dead} + \Delta m \quad (24).$$

These experiments were repeated from August to September of 1988 at the Faculty for Natural Science and Technology, Ljubljana. Two Mettler Zurich scales, type H2oT were used in the measurements. Identical results were obtained.

In another experiment, a test-tube was filled with 70 grams of live Californian worms and a small test-tube was filled with 0.25 ml of 36% water solution of formaldehyde. The control test tube is containing 70 ml of distilled water with a small test tube of formaldehyde inside. Both the test tubes were welded, wiped clean with 70% ethanol, and put into the weighing chamber of the balance.

Figure 4: control test tube (left) and experimental test tube

Approximately, one hour was allowed for acclimatization. Later both test-tubes were measured three times at intervals of five minutes. Then the test tubes were turned upside down to spill the solution of formaldehyde and again they were measured seven times at intervals of fifteen minutes. The weight of the test-tube with the worms was found to have increased in the first 3 minutes after the poisoning on average for an average weight of 60 micrograms and it then went down. Fifteen minutes after poisoning, the weight diminished on average by 93.6 micrograms.

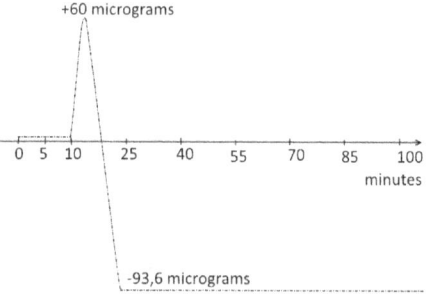

Figure 5: Mass increases and mass diminishing by 70 grams of Californian worms

This last experiment was repeated twelve times. The standard deviation goes to 16 micro-grams. The pressure in both test tubes was one atmosphere for the entire duration of the experiment as well as the temperature remaining unchanged. Neither the pressure nor the temperature could have been the cause for the change in the weight. Experiments are preliminary and need to be repeated at least in two different laboratories. The best would be using the balance "Mettler – Toledo Mass comparator AX107H" which can measure difference between two masses of 111 grams on 0,1 microgram precisely. This balance has two measuring vessels on which we put test tube and control test tube. We measure the difference between living and dead worms directly on the basis of the difference between masses of test tube and control test tube after poisoning of the worms. We can also put in both test tubes living worms and bring worms in one test tube to death with the appropriate radiation as for example gamma radiation. We do not encourage researchers to use higher developed animals in this experiment.

In 1997, one of the authors (A. Sorli) published the results of the experiments in the 'Newsletter' nr. 18-19 of Monterey Institute for Study of Alternative Healing Arts, California. On March 3rd 1998, Dr. Shiuji Inomata from Japan informed the editor (S. Savva) that Dr. Kaoru Kavada got similar results using rats as the experimental organism, again in a closed system.

Symbol Δm in equation (4) cannot denote ordinary 3D matter. In this case, so much energy would be released by the dead of worms that entire faculty would be destroyed. What is happening by the dead of worms is that the structure of subtle 4D biofield is falling apart and this is minimally increasing the energy density of 4DSS which was measured as the decreasing of the weight of the test-tube with dead worms.

The 'Biofield' is made out of 4DSS energy and additionally diminishes energy density of 4DSS in the living organism which results as the minimal increase of the weight. When biofield falls apart the weight is getting less.

Similar result happens by the relativistic particle which additionally absorbs the energy of 4DSS and so its energy and mass are increasing. By diminishing its speed particle will lose additional energy and additional mass (Sorli et al., 2018c). This additional energy of relativistic particle is its kinetic energy. When physical body or massive particle is back at rest the part of kinetic energy is released in the form of warmth and light. The other part is transformed back into the primordial energy of 4DSS. Similarly, when organism dies, the biofield is transformed into increased bio-photon emission. Slawinski research confirms at the dead time bio-photon emission is increased by 10-100 times (Slawinski, 2002). The other part of biofield is transformed back into the primordial energy of 4DSS.

living mass = dead mass
 + biophoton emission
 + dissolution back
 into 4DSS

relativistic mass = rest mass
 + photon emission
 + dissolutin back
 into 4DSS

Figure 6: Mass difference by relativistic mass and by living mass

Repetition of this experiment will give more data regarding the hypothesis that increased bio-photon emission is related to unexplainable phenomena at the time of death: "Considering the results from few recent scientific investigations, here we propose that specific mechanisms involving biophoton emission could probably be related to unexplainable phenomenon surrounding the moment of death" (Shashi, 2016). In this article we present that these "unexplainable phenomenon" is the separating of biofield structure from the physical body which results as the minimal diminishing of mass.

In Advanced Relativity, the bio-field has nine subtle layers, similar to yoga (Sorli at al., 2017). With only the acknowledgment of these subtle layers of life, traditional Western medicine will acquire the necessary tools to increase human health. Today Western medicine's focus is only on healing and preventing diseases and so has no remarkable success; in the West, diseases are constantly increasing. Western medicine has to focus on the health increase also, which has its basis in the subtle layers of bio-field. When subtle layers of bio-field are in tune with nature and consciousness, the human being is generally become healthy. Western medicine is focused only on the healing of the molecular layer of the human being, of which the proper functioning depends on the higher subtly layers of bio-field. The harmony and disharmony of subtle levels refers to the health and the illness (which results as the sleekness on 3D body level) respectively. This article is just a humble step for opening the eyes of Western science, for the integration of Western and Eastern medicine which will have immense benefits for entire humanity.

In this presented model of UFT, the photon is the fundamental element of the universe. Photon is "mass-less" in the sense that it does not have rest mass. Photon has energy and correspondent mass:

$$m = \frac{h \cdot v}{c^2} \quad (25),$$

where m is the mass of the photon, v is its frequency, h is Planck constant and c is speed of light (Sorli at al., 2018c). Photon is not mass-less in the sense that it does not have mass; this would be against "mass-energy equivalence principle". Photon has energy and so mass, it does not have "rest mass".

In today physics, there is also no clear difference between "mass" and "rest mass". Prevalent idea is that "mass" and "rest mass" are two names for the same phenomena. For example "mass" of the proton is its "rest mass". This is not the right imagination. "Rest mass" of the proton is its mass (as the amount of energy) when proton is at rest respectively to the 4DSS in which exists. Inertial mass of the proton at rest has origin in the diminished energy density of 4DSS (see Figure 2). When photon is accelerated (as for example in CERN accelerator) it gains its "relativistic mass" and its correspondent "relativistic energy". When two protons collide, which is extremely rare (once in milliard collisions), Higgs boson is released. Higgs boson is the artificial "sparkle of energy" with the extremely short life time and has no existence in physical reality on its own. That's why Higgs mechanism is from the sight of epistemology and methodology, which are extremely unstable (*Sorli at al., 2018c*).

The value of proton "relativistic" mass is equal for all observers independently on their speed of motion in 4DSS. Relativistic mass is defined exclusively by the speed of the proton in 4DSS which is the absolute frame of reference (*Sorli, 2018b*). The imagination: 'because of the different velocity of the observers, the relativistic mass of the proton is different' is not adequate. The mechanism why this should be so, was never explained and was introduced as "ad hoc". The fact is that in the relativistic proton energy of 4DSS is additionally absorbed because of the "dragging effect" and is valid for all observers (*Sorli, 2018b*).

The whole idea: 'the relative velocities of the observers define the rate of clock that running in 4DSS', is also not right. GPS proves that rates of all clocks on the satellites are valid for all observers. Rate of clocks depends exclusively on the energy density of 4DSS. Going away from the given stellar object, the rate of clocks is increasing. In interstellar space, the rate of clocks is also at maximum, where energy density of

4DSS is at maximum (*Sorli, 2018b*). In general, we can say that "relativity" depends only on the relative energy density of 4DSS and is valid for all observers.

Kinetic and Relativistic energy

When a given physical object or massive particle is at rest in 4D space there is no energy of space structured in the object. When object start moving the energy exchange happens, object absorbs energy of space which is its kinetic energy and by high velocities its relativistic energy. Motion of the object or massive relativistic particle additionally diminishes energy density of the space

$$\rho_{SE} = \rho_{PE} - \frac{mc^2}{V} - \frac{mv^2}{2V} \quad (26).$$

Formula (26) is for kinetic energy of massive object. For the relativistic massive particle formula for diminished energy density of space is following:

$$\rho_{SE} = \rho_{PE} - \frac{m_0 c^2}{V} - \frac{\gamma \cdot m_0 c^2}{V} \quad (27),$$

where m_0 is rest mass and γ is Lorentz factor.
Formula (27) we can develop:

$$\rho_{SE} = \rho_{PE} - \frac{m_0 c^2 \cdot (1-\gamma)}{V} \quad (28).$$

The famous Einstein "thought experiment with the elevator has new explanation: when the rocket is accelerating for $9{,}8 ms^2$ the kinetic energy of the rocket decreases energy density of the space so that it becomes equal to the energy density on the Earth surface, see equation (26) above.

Figure 5: Accelerating elevator

Technological applications of 4DSS

The 4DSS model is the unified field of the universe. Electromagnetism is an excitation of 4DSS; inertia and mass have origin in diminished energy density of 4DSS, caused by the presence of a given massive physical object. In Standard Model, there is no vision of unification between QED, gravity field and Higgs field, which actually have not a primary physical reality. QED is the basis for our computers, Einstein's Relativity is the basis for the satellites of GPS system and Higgs mechanism will never have technological application because it does not correspond to the real world.

The variable energy density of 4DSS is the basis for development of antigravity technology. When we will be able to increase energy density of 4DSS with the technical device, definitely, we can make antigravity. Calculations confirm that infinitesimally small increase of energy density is needed in order to gain antigravity. Very small decreases of energy density of 4DSS are causing gravity and a tiny increase will give us antigravity (Sorli, 2018b).

Casimir effect is the effect where pressure of 4DSS on the outer side of the plates is stronger than the pressure on the inner side of the plates. Experiments should be done with the different materials in order to see if materials with higher density (for example gold or osmium) increase Casimir effect. This would mean that given materials are less permeable for the energy of 4DSS. On that their property we could build the antigravity turbine whose rotation will cause minimal displacement of 4DSS energy which will cause minimal increasing of energy density of 4DSS.

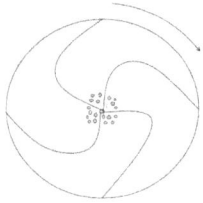

Figure 7: Antigravity Turbine

The other technological application is free energy technology, already developed by Nicola Tesla. It seems that 4DSS is electrically diametrically charged where stellar objects are present, namely, 4DSS above the Earth surface is electrically negatively charged and under the surface is electrically positively charged. On this basis Tesla tower was inducing electrical current directly from 4DSS. Continuation of his research seems the only realisable way to develop free energy technology.

Conclusions

Average mind is the main obstacle for the development of physics. Scientific thinking of 20th century science needs rejuvenation in order to embrace entire existence and not limit its methodology on scientific propositions for which we see are not correct. It is not mistake to think sometimes wrong. The mistake is not to realize that the thinking was wrong and keeping with the old way despite of new scientific discoveries, which require a paradigm shift. Standard Model belongs to the history of physics and requires appropriate extensions. We need a beyond Standard Model physics. We need a new fresh view of physical modelling, in order to progress physics and develop its bijective correspondence with the entire reality. This article is a modest attempt in this direction.

References

Ali A.F. and Das S., "Cosmology from quantum potential", *Physics Letters B* 741, 276-279 (2015).

Beverly Rubik, David Muehsam, Richard Hammerschlag, Shamini Jain, Biofield Science and Healing: History, Terminology, and Concepts, Glob Adv Health Med. 2015 Nov; 4(Suppl): 8–14.

Haramein N., Quantum gravity and the holographic mass. Physical Review & Research International 2013; 3(4): 270-92.

Ofer Firstenberg, Thibault Peyronel, Qi-Yu Liang, Alexey V. Gorshkov, Mikhail D. Lukin, Vladan Vuletić, Attractive photons in a quantum nonlinear medium, *Nature* volume 502, pages 71–75 (03 October 2013)

Fiscaletti D. and Sorli A., Bijective Epistemology and space-time, Foundations of Science, 20, 4, 387-398 (2015).

Fiscaletti D. and Sorli A., "About a three-dimensional quantum vacuum as the ultimate origin of gravity, electromagnetic field, dark energy ... and quantum behaviour", *Ukrainian Journal of Physics* 61, 5, 413-431 (2016).

Fiscaletti D. and Sorli A., "Quantum relativity and quantum vacuum", Ukrainian Journal of Physics, 63, 7, 623-644 (2018).

Korotkov Konstantin, Scientific basis of GDV method, https://www.korotkov.eu/scientific-basis/ (2018)

Slawinski J., Photon Emission from Perturbed and Dying Organisms: Biomedical Perspectives, Research in Complementary Medicine, Vol 12, Num 9 (2002)

Shashi Kiran Reddy, Could 'Biophoton Emission' be the Reason for Mechanical Malfunctioning at the

Moment of Death?, NeuroQuantology, Volume 14, Issue 4, page: 806-809, (2016)

Sorli A., Bijective Analysis of Physical Equations and Physical Models, NeuroQuantology 2018a, 16(6):

Sorli A., Advanced Relativity: Reintroduction of absolute Frame of Reference, NeuroQuantology 2018b,

Sorli A., Dobnikar U., Patro S.K.,Mageshwaran M. ,Fiscaletti D., Euclidean-Planck Metrics of Space, Particle Physics and Cosmology, NeuroQuantology 2018c, 16(4): 18-25

Sorli A., Dobnikar U, Fiscaletti D, Koroli V. Advanced relativity: multidimensionality of consciousness and mind, origin of life, psi phenomena. NeuroQuantology 2017; 15(2): 109-17.

Sorli A., The Additional Mass of Life, *Journal of Theoretics*, Vol.4-2, (2002) http://www.journaloftheoretics.com/Articles/4-2/Sorli-final.htm

Sorli A., Patro S.K. Bijective Physics, ISBN 9781721801725, AMAZON, (2018)

Reader notes: